*f***P**

DO IT SMART

Seven Rules for Superior
Information Technology Performance

ROLF-DIETER KEMPIS
JÜRGEN RINGBECK
RALF AUGUSTIN
GÜNTER BULK
CHRISTOPHER HÖFENER
BERTHOLD TRENKEL-BÖGLE

THE FREE PRESS

THE FREE PRESS
A Division of Simon & Schuster, Inc.
1230 Avenue of the Americas
New York, NY 10020

First Free Press Edition 1999
Published by arrangement with Wirtschaftsverlag C. Ueberreuter GmbH
Originally published in Germany as *Do IT smart. Chefsache Informationstechnologie—Auf der Suche nach Effektivität*

THE FREE PRESS and colophon are trademarks
of Simon & Schuster, Inc.

Designed by Michael Mendelsohn of MM Design 2000, Inc.

Manufactured in the United States of America

10 9 8 7 6 5 4 3 2 1

Library of Congress Cataloging-in-Publication Data

Do IT smart: seven rules for superior information technology
performance / Rolf-Dieter Kempis . . . [et al.].
p. cm.
Originally published in German under the same title.
Wirtschaftsverlag Carl Ueberreuter GmbH, Frankfurt a. Main, 1998
1. Information technology—Management. 2. Strategic planning.
3. Industrial management. I. Kempis, Rolf-Dieter.
HD30.2.D6 1999
658.4'038—dc21 99-21275
CIP

ISBN 0-684-86425-8

CONTENTS

FOREWORD

In the age of the Internet, the use of sophisticated information and communication technology in industry has become second nature. But what contribution does this wealth of hardware, software, and networks actually make to a company's bottom line? How far do the benefits of IT depend on its functionality? What impact do factors such as quality of the infrastructure or the design of individual business processes have? And what level of expenditure on IT is reasonable?

My colleagues Dr. Rolf-Dieter Kempis and Dr. Jürgen Ringbeck decided to investigate these questions with a team of McKinsey consultants and staff from the Institute of Production Engineering and Machine Tools at Darmstadt University of Technology, focusing on the manufacturing industry. In the course of a 15-month research project, they examined leading international companies and supplemented these analyses by countless discussions with top managers and experts.

A key finding of the survey is that smarter solutions to the question of *what* IT is used for—IT effectiveness—make a much greater contribution to corporate success than concentrating on the *how* of IT support, IT efficiency. Leveraging both virtues simultaneously is the best strategy of all, of course. But only around 25 percent of all companies analyzed demonstrated both above-average IT efficiency and effectiveness, earning the title "IT stars." Even within this group, there were differences between the "simple" stars and the best of the best. But the stars provide outstanding benchmarks in many respects.

One of the main breakthroughs of this study was the development and application of a methodology for measuring IT effectiveness. Now a metric is finally available for a variable that every manager sensed intuitively, but could not express in figures. The quality of IT in individual core processes in terms of its functionality, availability, and utilization can be used to quantify IT effectiveness for specific processes. Whether for product development, marketing, sales, order processing, or service, numerous indicators were combined to make an effectiveness index—a highly ambitious empirical and analytical achievement.

The decisive link in the chain of cause and effect between IT effectiveness and corporate success is process performance in the business processes supported by IT. The prerequisite for improving these processes is their design or redesign. Many companies have recognized this connection, performing reengineering projects in an attempt to lay the foundations for more effective IT support, which—if implemented professionally—is also more efficient, i.e., cheaper and faster. The special feature of this book is that it crystallizes the findings of the McKinsey survey into seven rules for successful IT management in every company.

Dr. Kempis, Dr. Ringbeck, and their team do not limit their perspective to the survey results. They also consider the future perspectives of IT. Instead of succumbing to the usual IT euphoria, they draw sober conclusions from practical experience and observable trends. Courage to innovate will continue to reap rewards, but understanding and controlling the new information and communication technologies will become even more important than ever before. In view of the ever increasing performance of IT and its shrinking innovation cycles, this means the ability to move fast, and to concentrate on the essential: the systematic integration and standardization of a company's different information technologies.

New webs of integration will develop both internally and beyond a company's borders. Learning communities within companies will open up their intranet solutions even further, extending them from R&D through production to marketing, sales, and service, for example. These communities will also successively integrate selected external business partners. A healthy dose of skepticism is needed toward prophecies of an Internet-dominated future, but communication networks such as the "Automotive Network eXchange" (ANX) of Chrysler, Ford, and GM are witness to the enormous productivity reserves that can be captured via the smart application of these new technologies.

Integration of the different IT systems must be closely interconnected to their standardization. Agreement on uniform, binding communication standards and data (exchange) formats is just as important as the growing use of integrated standard software until "brainware" standards have been established that allow the coordination of companies in virtual enterprises. Standardization will then become a guarantee of flexibility.

The design and guidance of such all-encompassing integration and standardization initiatives is a top management affair. The more management attention is focused on IT issues of this kind, the more successful a company will be.

Detlev J. Hoch
Director, McKinsey & Company, Inc.

ACKNOWLEDGMENTS

This book could not have been written without the assistance of colleagues both within the firm and beyond. We particularly wish to thank Professor Herbert Schulz and Markus Haag from the Institute of Production Engineering and Machine Tools of the Darmstadt University of Technology, who supported us throughout our work with valuable advice and constructive feedback.

We would also like to thank many colleagues at McKinsey for their professional support; among these are Hans-Georg Frischkorn, Carolin Elger, Carsten von der Ohe, Andreas Cornet, Bengt Starke, Peter Wolpert, Georg Klymiuk, Torsten Oltmanns, Dorte Landwehr, Michaela Schneider, Feridun Keskin, Judith Burmann, Ursula Grewing, Sabine Kröll, and Gilian Crowther.

INTRODUCTION

Over the past two decades, almost no other subject has been discussed in the manufacturing sector as passionately as the use of IT. IT enthusiasts paint the opportunities for IT in glowing colors; they wax poetic about paperless offices, fully automated plants, and virtual enterprises. The skeptics respond with sarcasm: "There are many different ways to ruin a company. Speculation is the fastest, IT the most reliable."

Quotations from survey interviewees quickly indicated how divided the camps are, from top management through to operations:

- *R&D.* "Our new engineering information system will offer on-line access to all patent databases. Users can learn to operate it in one day using a CAD interface," say the supporters. Using their IT, R&D has created so many variants that we can't handle them anymore," production complains.
- *Sales.* "The product configurator has significantly improved our customer service. It has reduced the number of special requests enormously," argues the one side. The other: "Our sales information system is so inflexible that it's hampering the vital restructuring of order processing."
- *Purchasing.* "For the first time, our integrated materials management system enables us to evaluate suppliers comprehensively on a daily basis, analyze make-or-buy decisions reliably, and optimize logistics planning." This is mirrored by the opposite opinion: "In our sector of industry, IT support for purchasing doesn't make any sense. Interfaces with our suppliers are so complex—excellent personal staff contacts are all that count."
- *Production.* "The new integrated standard software will improve our production stability, boost our delivery reliability, and increase stock turnover tenfold," declares an IT protagonist, while an IT opponent replies: "Our production planners use their entire energy trying to outwit the enterprise resource planning system. It is totally inappropriate for a manufacturer of product variants."

- *Administration.* “Our extremely flexible executive information system is based on the data warehouse concept. It allows us to analyze problems systematically on screen in precise detail. This improves decision making—and saves a mint on paper!” A colleague’s reaction is the complete opposite: “Interfaces between the financial accounting systems of our plants are weak—every transaction has to be checked again manually.”

We constantly encountered these conflicting attitudes in the course of our interviews. Why is there such a gap between aspiration and reality across all lines of business in the manufacturing sector? Before our survey, no systematic correlation had been shown between the use of sophisticated hardware and software systems and the operating results of companies, although isolated cases have time and again demonstrated the cost and competitive advantages that can be captured from professional use of IT.

Our research project, Benefits of IT in the Manufacturing Industry, was designed to close this gap in the field of IT research.

Together with the Institute of Production Engineering and Machine Tools at Darmstadt University of Technology, we conducted in-depth interviews with top managers of over seventy leading industrial companies in the manufacturing sector from Europe, the United States, and Asia. The spectrum of industries ranged from component manufacturers, automotive suppliers, mechanical engineering companies, and electronics firms to companies in the process industry. We also interviewed experts at another thirty companies on specific themes, such as the use of integrated standard software, outsourcing, rapid prototyping, and product configurators, and held discussions with selected suppliers of these IT products and services.

We developed a new, strictly empirical methodology for measuring the impact of IT on corporate success. Analysis of the survey results revealed that although IT efficiency is important, IT effectiveness makes a particularly powerful contribution to a company’s bottom line. We found that highly effective IT organizations seem to share an understanding of seven key principles.

This book describes those seven rules for superior IT performance. It is addressed to managers in the manufacturing sector who have to make fundamental decisions on IT strategy and investment, and to executives and users from departments and functions who wish to learn more about the strategic and operational benefits of IT. We hope the factors for success described in this book will help readers tap the enormous productivity reserves that can be opened up by smart IT management.

Part I

SEVEN RULES MAKE THE DIFFERENCE

How do you define good IT performance? A programmer or systems analyst might measure it using MB/s or MIPS. But what does that really mean to the user? From a user perspective, IT performance reflects both the efficiency and effectiveness of IT management and systems. Throughout this book we use the following definitions:

- *IT efficiency:* IT cost as a percentage of revenues, plus project management performance against schedule and budget
- *IT effectiveness:* The availability, functionality, and utilization rates of IT applications for each core business process

The impact of IT efficiency on corporate success (i.e., profitability and growth) can be measured directly (see Exhibit I–1). However, the impact of IT effectiveness on corporate success can only be measured indirectly through process performance in product development, in the operational core processes—marketing and sales, materials and logistics management, manufacturing and service—and in administration.

For the empirical research, we developed a comprehensive performance measurement system that mirrors this correlation and delivers the basis for the analyses. (See Appendixes A and B for details on the approach we used in our empirical research, scope of the survey, structure of the companies analyzed, and the methodology of our performance measurement system.)

The system measures the benefits of IT using four indicators of corporate success: return on sales, change in return on sales, change in sales,

Exhibit I–1
Measurement of IT Performance

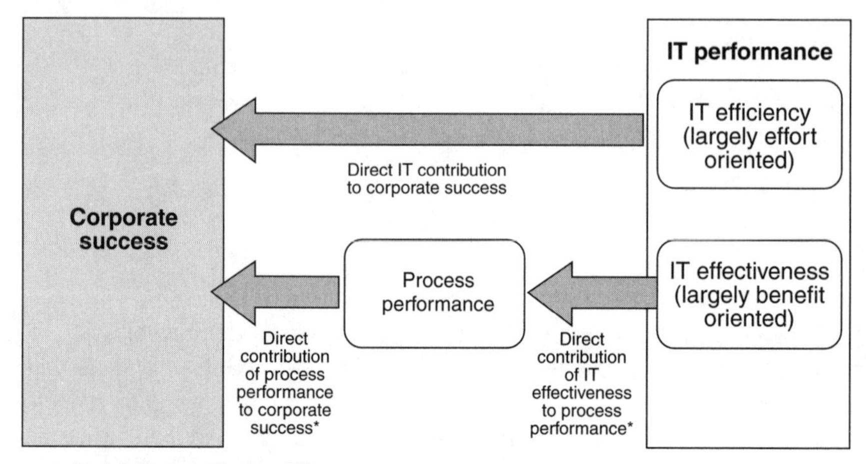

* only indirect contribution of IT to corporate success
Source: McKinsey & Company, Inc.

and change in market share—all the indicators that are commonly used to measure bottom-line corporate performance.

THE FOUR IT CULTURES

The companies surveyed fell into four categories depending on their IT efficiency and effectiveness. We called these categories the "four IT cultures":

- IT stars
- Big IT spenders
- Cautious IT spenders
- IT laggards

Exhibit I–2 illustrates the spread we found.

Exhibit I–2
IT Performance: Four IT Cultures

Only a few companies are IT stars

Percentage of companies surveyed

IT support of core business processes
- Functionality
- Availability
- Utilization rate

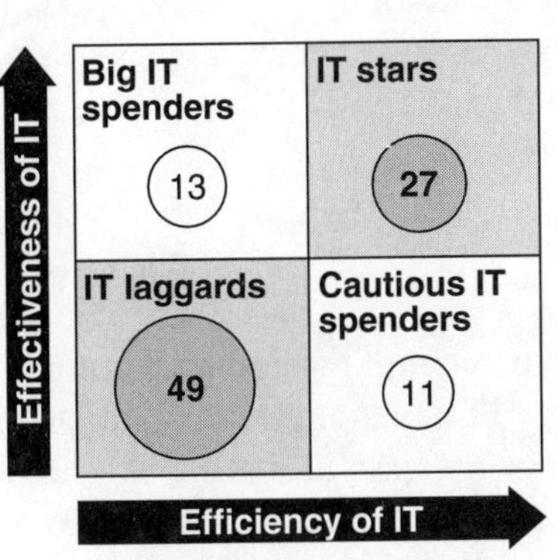

- IT cost as a share of sales (adjusted by industry segment)
- IT project management (meeting deadlines and budget)

Source: McKinsey & Company, Inc.

Only 27 percent of the companies surveyed are IT stars with excellent IT performance

Our research showed that only 27 percent of the companies analyzed have both effective and efficient IT. Forty-nine percent of companies achieve disappointing functionality, systems and data availability, as well as low utilization rates at high cost.

IT STARS

At comparatively low IT costs, IT stars achieve very good IT effectiveness. Users of IT systems in these companies have good IT know-how, and the IT department sees itself as a process-oriented internal service provider. We could classify only 27 percent of the companies surveyed as IT stars.

BIG IT SPENDERS

Thirteen percent of the companies surveyed achieved above-average IT effectiveness—at the price of above-average effort and expenditure. They give the go-ahead to almost any application that looks as if it might be useful, regardless of the expense. Often they favor expensive proprietary software solutions rather than cheaper standard software, imagining this will secure them a competitive advantage.

CAUTIOUS IT SPENDERS

Eleven percent of the companies analyzed are efficiency oriented and control their IT expenditure via tight budgets. This prevents spiraling IT costs but does not provide sufficient support for these companies' business processes. Cautious spenders come nowhere near exploiting the full potential of IT technology. The reason is not only their prime focus on cost optimization (often due to acute business problems) but also lack of commitment on the part of their IT managers. IT departments at cautious spenders largely play a passive role instead of acting as change integrators.

IT LAGGARDS

Laggards make up the biggest category, at 49 percent. In spite of high IT costs, these companies derive little benefit from their IT. These companies have underdeveloped organization structures with a lack of service orientation and poor project management. Often laggards have only a vague conception of potential IT performance, and their IT costs and follow-up expenditure are intransparent.

HOW THE FOUR IT CULTURES PERFORM

How could IT laggards benefit from bringing their IT management up to scratch? And how do greater IT effectiveness and efficiency really affect a company's bottom line? To answer these questions, we used a financial indicator to score the companies surveyed, consisting of four criteria: return on sales, change in return on sales, change in sales, and change in market share (see Exhibits I–3 and B–1). The corporate success of each company was rated on a scale ranging from 0 (poor) to 100 (excellent). On the basis of this financial indicator, it emerged that IT stars perform substantially better than IT laggards and cautious IT spenders (see Exhibit I–3).

It is noteworthy that at constant levels of low effectiveness, an increase in efficiency merely helps to lower IT costs. In terms of corporate success, cautious spenders rated insignificantly better than laggards on average.

The effects of improving IT effectiveness are much greater. Big IT spenders and IT stars demonstrate much better results along the indicators of corporate success than IT laggards and cautious IT spenders, which have greatly inferior IT effectiveness.[1] An improvement in efficiency has less bottom-line impact than an effectiveness-oriented approach aimed at achieving competitive differentiation via smart IT support of processes (see Exhibit I–4).

What explains the difference in leverage of IT effectiveness versus efficiency? To find out, we need to examine the underlying criteria of the financial indicator (see Exhibit I–5).

Exhibit I–3
Financial Indicator of the Four IT Cultures

More effective IT correlates positively with corporate success

Indicator of corporate success: Index 0–100 (poor–excellent)

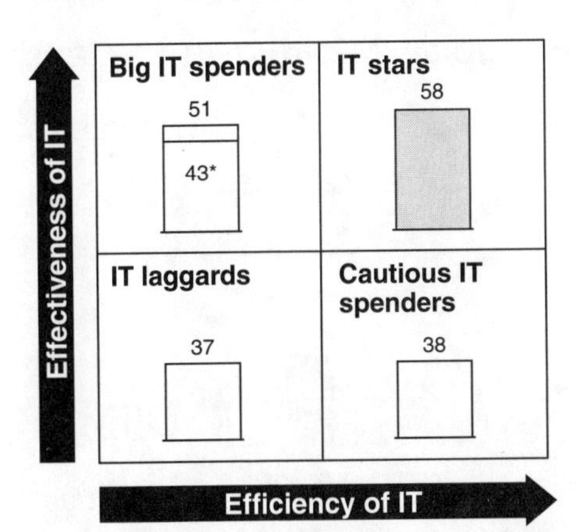

Indicator of corporate success comprises

- Return on sales
- Change in return on sales
- Change in sales
- Change in market share

* Excluding companies with dominant market positions due to patents or other reasons, no companies with dominant market positions existing in the other matrix segments

Source: McKinsey & Company, Inc.

[1]The relatively high value of the financial indicator of big IT spenders must be viewed with caution because several companies have dominant market positions due to patents or other reasons, thus increasing the averages of the category.

Exhibit I–4
Key Success Factors of IT Management

While efficiency is important, effectiveness makes a particularly powerful contribution to corporate success

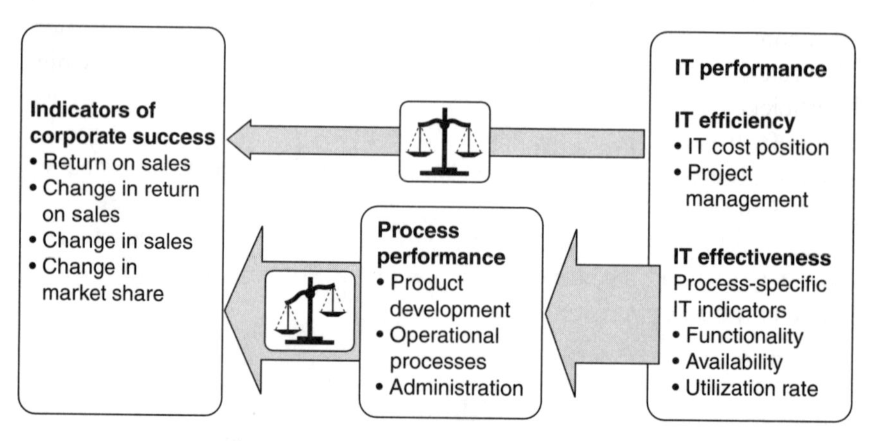

Source: McKinsey & Company, Inc.

Exhibit I–5
Underlying Criteria of the Four IT Cultures

IT stars achieve above-average and rising returns, and grow faster than their markets

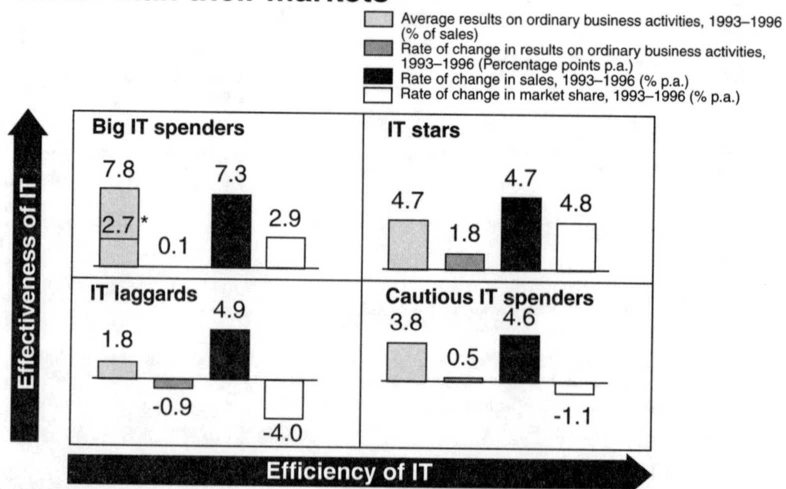

* Excluding companies with dominant market positions due to patents and other reasons; no dominant market positions existing in other matrix segments; no major influence of dominant market position on change rates.

Source: McKinsey & Company, Inc.

- **IT efficiency increases return on sales.**
  On average, companies with higher IT efficiency achieved a higher return on sales than companies with lower IT efficiency in the period 1993–1996 examined. Regardless of whether the company has a low or high IT effectiveness rating, this difference is stable and amounts to two percentage points.[2] There was also a strong self-reinforcing effect: in companies with high IT efficiency in 1993–1996, the rate of change in return on sales was clearly positive, while in companies with low IT efficiency, the change was negative or only marginally positive.
- **IT effectiveness stimulates revenue growth.**
  High IT effectiveness is a driver of revenue growth. From 1993 to 1996, the average sales growth of companies with high IT effectiveness was 7.3 and 7.4 percentage points, respectively, per year. Companies with low IT effectiveness only achieved in the same period a growth rate of 4.9 and 4.6 percentage points, respectively, per year in the same period. This relationship can be seen even more clearly in the relative change in market share defined as nominal change in sales minus change in market volume. Companies with high IT effectiveness increased their market share, whereas companies with low IT effectiveness actually lost market share. This was true almost regardless of whether the companies use IT with high or low efficiency.

What is astonishing is that there do not seem to be any trade-offs. IT stars demonstrate that you can achieve high IT effectiveness without sacrificing IT efficiency. This becomes even clearer if you analyze the results of IT stars in detail (see Exhibit I–6). Even among this group, however, there is still significant improvement potential.

Even IT stars have significant improvement potential

With regard to IT effectiveness, the best of the best in product development, operational core processes, and administration achieved results that were about 10 to 25 percent above the average of the relevant category. With IT efficiency, the best of the best were in a special situation: measured by the broad concept of IT costs that we used, they spent only 1.3 percent of their revenues on IT, while IT stars spent on average 1.7 percent. This superiority was also sharply reflected in project management. Deviations from estimated project costs and duration were about 5 percent, while IT stars on average achieved only 10 to 15 percent. Although IT stars clearly stand out from the other IT cultures, many stars still have significant room for optimiza-

[2]Companies with dominant market positions were not taken into account.

Exhibit I–6
Leveraging IT to the Full

Even IT stars can distinctly improve their IT performance

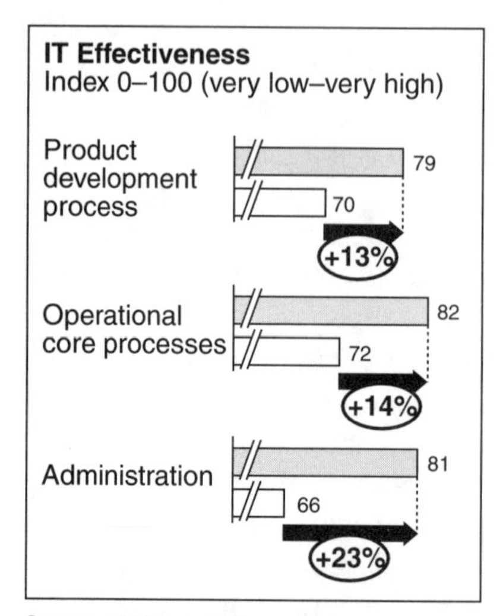

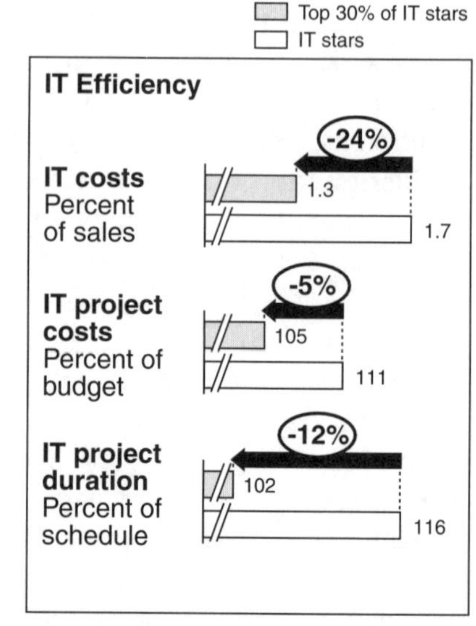

Source: McKinsey & Company, Inc.

tion. This means not the IT as such but its superior use sets the best of the best apart from other IT stars.

These results show that excellent information management is an essential prerequisite for success. For top management it is just as important as the other more familiar drivers of strategic management, such as complexity management and market orientation. Few companies can afford to behave like big IT spenders and neglect costs. This makes it crucial to scrutinize the spectrum of IT services available with clear objectives in mind. There is no magic wand for conjuring up the right IT management: every company will have to find its own way. But companies can gain a great deal of inspiration from the best practices of IT stars.

SEVEN RULES FOR MANAGING SUPERIOR IT PERFORMANCE—AN OVERVIEW

What are the secrets of the IT stars' success? In the course of our analysis, we found a number of principles that they all seem to be guided by: the seven rules for managing superior IT performance.

1. Make IT a priority in product development.
2. Integrate IT into marketing, sales, and service.
3. Use IT selectively to integrate order processing across the company.
4. Shift the focus of IT in administration to business planning and management development.
5. Make IT a top management affair.
6. Create a customer-oriented IT service network.
7. Introduce integrated standard software on a fast-follower basis—but redesign the business first.

RULE 1: MAKE IT A PRIORITY IN PRODUCT DEVELOPMENT

Companies with superior IT management tend to be better at developing products and achieve better results with smaller budgets. As best practice moves beyond heavy investment in CAD software, two areas are emerging as the future focus of first-class product development.

The first is **simulation and calculation software.** The use of software to perform simulations and calculations in product development is nothing new. But successful developers boost development output by using them earlier and more intensively. In our survey, approximately 50 percent of stars claimed to use IT to identify development errors such as design misspecifications quickly and reliably, compared to only 25 percent of laggards.

The application of simulation software to modeling is already cutting development time and cost, and its future holds even greater promise. Through rapid prototyping rather than conventional tooling up, an auto parts manufacturer managed to reduce manufacturing costs on a preseries of rear lights by 60 percent *and* shorten delivery time by 75 percent.

The use of assembly and production simulation in earlier development phases fosters design for manufacturing and assembly (DFMA), thus reducing manufacturing costs.

The other area is **electronic integration of engineering data and tasks.** The integration of data (such as CAD data, simulation results, work schedules, and project status information) with tasks in the development process has much to offer. Key challenges in R&D management, such as reducing development time and broadening technical expertise, are forcing companies to integrate their own core processes and their development partnerships with other companies.

Successful developers are ahead in integrating product data management (PDM) with other applications: 50 to 70 percent of these companies have integrated information on calculation and simulation results and work plans into their PDM systems; for less successful developers, the figure is only 13 to 31 percent. Successful developers also use PDM to estimate product costs earlier in the product development process. The integration of engineering databases can serve as a springboard to faster, simpler communication with engineering partners.

Seeking to encourage superior DFMA—the design of products to optimize quality-to-cost trade-offs—more than 75 percent of successful developers have built up interfaces between direct numerical control (DNC) machinery management in the production area and CAD systems. Among less successful developers, only 40 percent have achieved this degree of integration. The story is much the same in relation to the integration of CAD and simulation applications in manufacturing and assembly, and with development partners.

A comparison of companies' goals for product development and the application of new technologies in the coming years reveals that the gap between more successful and less successful developers in the application of IT will continue to widen.

RULE 2: INTEGRATE IT INTO MARKETING, SALES, AND SERVICE

Traditionally sales is an area of low IT penetration. Management attempts to use management information systems to make the sales process more transparent have been met with skepticism from sales staff. Even so, master databases for customer files have become standard, as have order processing systems that permit faster and more consistent order transfers. In addition, over 60 percent of the sales staff surveyed are now equipped with laptop computers and mobile phones, and many companies provide product and price information in CD-ROM catalogues and other digital formats.

Companies whose sales performance is above average tend to use integrated standard software more intensively, allowing them to integrate sales data to a greater extent. Their sales information systems also provide direct access to data on capacity and production scheduling, so that the feasibility of any customer request can be checked before an order is placed. Some companies have gone further, allowing regular customers to enter their orders directly into their systems via electronic data interchange (EDI).

Most important, the survey suggests that the use of a new application software—a product configurator (electronic tools that simulate putting together a product from thousands of possible features)—will make an important contribution to business success in the future. During a sales pitch, staff will be able to show customers the full range of products, variants, and types; check on consistency and feasibility of the order; calculate the price of the product chosen; and agree on a realistic delivery date.

The advantages are obvious. Contracts for products that a company cannot manufacture would become a thing of the past, as would most delivery delays. Jobs could be scheduled on the spot and products engineered at estimated cost. By using a configurator, one manufacturer was able to replace a full 90 percent of individual customer requests with a set of precosted variants.

When it comes to service, companies with excellent performance stand out because of the applications (such as diagnostic systems) they employ and the customer and product data they have available. Partly as a result of this broader range of applications, companies with the best service performance use their IT systems twice as intensively in the field as companies with poor service performance.

RULE 3: USE IT SELECTIVELY TO INTEGRATE ORDER PROCESSING ACROSS THE COMPANY

Purchasing, materials management, and production planning have long been key areas for IT applications. By monitoring materials and comparing stock levels with production requirements, IT can help organize orders efficiently. Information systems are also indispensable for planning production jobs down to the calculation of daily job lists for individual production areas.

Despite these advances, the factory of the future has yet to become reality. Many companies feel their production planning and control systems are ineffective, inappropriate, and brittle. Companies with complex production planning programs and small batch sizes complain that their production planning software does not cater for their needs. Even manufacturers of standard products find their systems inflexible when orders change.

Successful manufacturers handle these difficulties by knowing when IT works and when it does not. Surprisingly, although IT plays an important role in other core operating processes, it seldom offers much help with production planning at a detailed level. Almost 80 percent of companies that excel at order processing dispense with IT-supported control at the shop floor level, instead using simple organizational processes such as Kanban (a streamlining plan that drives production of materials on demand and maintains inventory at the point of use). Their importance as a substitute for IT increases with the complexity of the production process.

Successful developers also limit their customers' flexibility over scheduling orders: their freeze point for placing orders in their materials requirements planning system is much earlier, and once it is reached, they change orders less often to avoid disrupting production schedules.

IT has an important contribution to make to purchasing and logistics. IT-supported inventory management is standard; what makes the difference is the monitoring of orders between a company and its customers and suppliers via EDI. Today EDI is heavily used to handle orders between automotive suppliers and original equipment manufacturers (OEMs). So far, the use of EDI has not been widespread among second-tier suppliers and in other branches. However, more intense use of EDI for customer order calls leads to much lower logistical costs.

RULE 4: SHIFT THE FOCUS OF IT IN ADMINISTRATION TO BUSINESS PLANNING AND MANAGEMENT DEVELOPMENT

Historically, administration—finance and accounting, controlling, and human resources—was the first part of a company to apply IT. All businesses now use IT extensively in this area, to the point that standard software seems to have largely exhausted its potential for efficiency gains.

True distinction can be achieved, however, if IT in administration can move beyond these conventional data processing tasks to cater for a new understanding of corporate planning, systematic management development, and effective knowledge management. One example is business planning. High-performing companies achieve twice the planning quality at half the costs, estimating their trend in return on sales with an average precision of ± 1.7 percentage points (versus ± 4.0 percentage points at less successful companies). High performers have integrated executive information systems—giving senior executives daily, weekly, or monthly updates on key dimensions of business performance—far more flexibly than their low-performing peers. Effective utilization of IT tools to mine the wealth of data available in base systems for short-term planning—and in special cases also for long-term planning—will be key in this shift of emphasis in administration.

In addition, the more advanced players use IT extensively in human resources for staff development and assignment planning, and place greater emphasis on generating employee profiles and job descriptions. Emerging work flow management systems may even create fresh administrative roles in IT to manage and organize knowledge within the organization.

RULE 5: MAKE IT A TOP MANAGEMENT AFFAIR

Information management must receive the attention of top management. At IT stars, top managers together spend about 45 hours per month on IT, compared with 20 hours for laggards. They devote time and energy to developing an IT strategy and get actively involved in the introduction of new systems. They play a critical role in defining projects and agreeing on measurable goals in specific business processes and technologies. Managers keep in touch with the progress of new projects, and spend more than twice as much time on IT training. Without this intimate involvement of top management in critical IT issues, information management rarely performs well.

RULE 6: CREATE A CUSTOMER-ORIENTED IT SERVICE NETWORK

Traditionally IT tasks are assigned to a central data processing department. However, leading-edge companies have reinvented their IT service structure. First, they create a network distributing IT service tasks throughout the company in dedicated units, committees, and project teams rather than assigning them to a single department. They set up an information management group to concentrate on IT planning and consulting, including process redesign. Its head chairs a steering committee—comprising top management, IT department heads, and members of a core IT user team—that makes decisions on key corporate IT issues. The core IT user team brings together leading IT users from all areas of the business, and defines user requirements in collaboration with the revamped IT department, which now concentrates on providing infrastructure and operating the systems. IT users also play a critical role in adding pivotal resources to software implementation projects.

Second, although third-party providers are a central part of IT service networks, IT stars are much more careful about what they outsource. Only 24 percent outsource operational IT services such as running the computer center; laggards do so at twice that rate. When IT stars farm out IT tasks, they do not do so for efficiency reasons but in order to increase effectiveness. They also rigorously monitor providers, assessing their performance through external benchmarks and meticulous goal setting, and pay close attention to the negotiation of contract extensions.

Third, IT stars are much more professional about their IT planning and control processes, relying on their information management group with its close links to top management. They understand their costs in greater detail; 53 percent of stars were able to produce detailed cost structure analyses, compared with just 7 percent of laggards. Less successful players often set IT budgets purely on the basis of historical data, while stars conduct zero-base budgeting, reassessing projected and current tasks every year.

Most IT stars study the market systematically to identify important innovations: 88 percent regularly invite vendors to make product presentations, as against 46 percent of laggards. More than one-third of IT stars assign staff to comprehensive market research, compared with 13 percent of laggards. Armed with up-to-date information, they plan for the long term: 65 percent formulate three- to five-year plans for hardware, software, and netware projects, as against 37 percent of laggards. Similarly, 50 percent of stars are planning major IT projects in the next three to five years, while only 25 percent of laggards have such projects planned that far ahead. Such detailed planning allows top performers to develop their IT capabilities deliberately and systematically instead of wasting energy on firefighting.

RULE 7: INTRODUCE INTEGRATED STANDARD SOFTWARE ON A FAST-FOLLOWER BASIS—BUT REDESIGN THE BUSINESS FIRST

In most situations it is better to use functionally integrated standard software than invest in proprietary solutions. Seventy-five percent of the implementation costs of IT stars are devoted to integrated standard software, compared to 42 percent for laggards. Pioneering proprietary software is the right strategy only when it produces a clear competitive advantage, as with some product-specific simulation software.

When to introduce integrated standard software is a key decision. A fast-follower strategy is the best bet. Companies should wait until early software bugs have been fixed and external consulting know-how has become available before making a commitment. But once new releases can offer greater functionality and user friendliness, companies should act fast.

Over 60 percent of IT stars follow this approach. Laggards tend to adopt more reactive strategies. Because they also continue to use large parts of the old software after new systems are introduced, they experience many more problems with compatibility.

The survey confirmed the importance of reengineering business processes *before* new systems are introduced rather than at the same time.

The sequential approach followed by over 40 percent of stars (but just 24 percent of laggards) reduces both project duration and costs by more than 50 percent. It enables stars to refine their selection criteria for software and gain a better understanding of how standard packages could be adapted to their requirements, thus avoiding the need for expensive custom programming.

IT stars apply to these projects the same planning and project control that distinguishes them in other areas. Although they resort to external implementation partners twice as often as laggards do, they tend to focus their partners' efforts on conceptual work and early pilots, and strive to involve their own staff—IT users as well as IT managers—more closely in the rollout.

Because they monitor IT milestones rigorously, stars stray from planned cost far less than do laggards. After new systems have been introduced, they monitor the utilization rate, evolution of users' knowledge, ongoing expenses, and improvements in business performance in order to pursue continuous improvement. By contrast, laggards often fall into a downward spiral. Frequent failure to meet objectives combined with dissatisfaction among IT users make it hard to win support for follow-up projects. Problems are swept under the carpet. Control over outsourcing lapses. The result is excessive spending for only mediocre results.

Part II

EFFECTIVE USE OF IT IN CORE BUSINESS PROCESSES

Our research indicates that professional IT management improves profitability and favors growth. To understand this relationship, we will look at the core business processes in more detail. We begin by explaining the definition of core business processes that we used and summarizing the impact on them of IT, and then discuss the processes individually.

DEFINITION OF CORE BUSINESS PROCESSES USED

The core business processes as we defined them in the manufacturing industry fall into three main categories:

- **Product development.** Includes all stages of the development process: developing a product concept, designing and detailing a prototype, testing, and market introduction.
- **Operational core processes** for the creation of goods and services. These break down into the following subprocesses: marketing and sales, order processing (materials and logistics management), manufacturing, and after-sales service.
- **Administration.** Corporate finance, accounting, controlling, and human resources management.

IMPACT OF IT

To analyze the impact of IT on core business process performance and corporate success, we focused on the following questions:

- How can core business process performance and its impact on corporate success be measured? What impact do IT-supported product development, operational core processes, and administration have on corporate success?
- How can the relationship between successful IT management and core business process performance be explained? Do companies that manage their IT better really have better process management?
- How do the best in class master the use of IT? What types of hardware and software do they use? What leading-edge solutions enable them to gain competitive advantage?

Operational core processes and product development performance are key drivers of financial success

Our research showed that operational core processes and product development performance are key drivers of financial success, IT stars clearly have superior process management, and they rely on smart IT applications and high utilization.

KEY DRIVERS OF FINANCIAL SUCCESS: OPERATIONAL CORE PROCESSES AND PRODUCT DEVELOPMENT PERFORMANCE

To prove this causal connection objectively for the first time, we evaluated time, quality, and costs of core business processes using our performance measurement system. As shown in the matrix in Exhibit II–1, companies were mapped by performance on product development and operational core processes. The impact of these processes on corporate success was measured by four criteria for each of the four quadrants.

High performance in operational core processes improves profitability; effective product development performance promotes growth

What the results show is that companies with higher performance in operational core processes achieve significantly better returns on sales, while companies with superior performance in product development achieve above-average increases in sales and market share. In contrast, there is no significant correlation between performance in administrative processes and corporate success, although more efficient administration improves cost effectiveness, thus contributing to financial results.

Our research reveals a strong correlation between operational excellence and superior IT management. However, excellence in one opera-

Exhibit II–1
Impact of Performance in Operational Core Processes and Product Development on Corporate Success

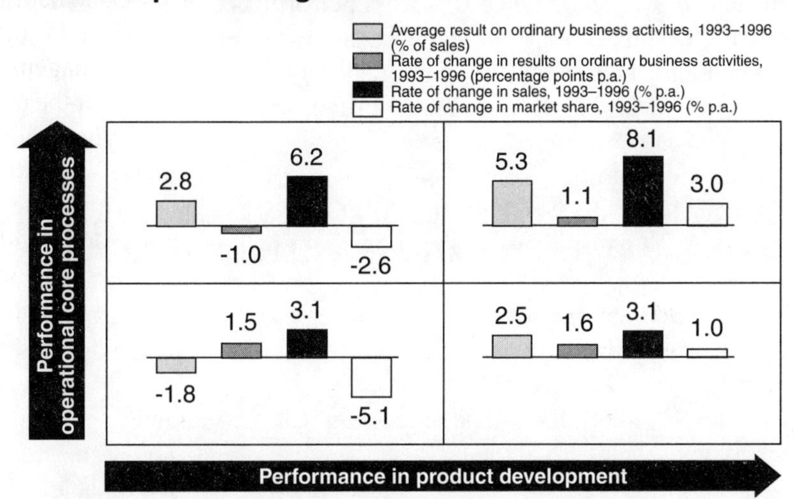

* Administrative processes have no demonstrable impact on corporate success.
Source: McKinsey & Company, Inc.

tional core process does not automatically mean excellence in all other operational core processes. For instance, best performance in sales is not necessarily linked to best performance in production, and vice versa.

To analyze the impact of IT on operational excellence, we distinguished between efficiency and effectiveness of IT support in different operational core processes. We found that most of the companies surveyed concentrated on improving process efficiency through IT support, thus letting valuable opportunities slip. The best of the best are far ahead: they have already started to leverage IT to improve process effectiveness—for example, by improving product quality and customer satisfaction.

IT STARS CLEARLY HAVE SUPERIOR PROCESS MANAGEMENT

Among the companies surveyed, all IT stars perform significantly better in product development, whereas 60 percent of IT laggards and 71 percent of cautious IT spenders are less successful performers in this area. This is hardly surprising considering the similarities between management of IT projects and R&D project management. Obviously, comparable management principles apply once a company has mastered these principles; it can transfer them with equal skill to either discipline.

Superior IT management correlates with excellent process management

Typically IT stars are good performers in operational core processes; 75 percent of IT stars have superior operational core process performance. And companies with good performance in operational core processes turn out to have better IT management. The research therefore shows a close empirical correlation between superior IT management and superior process management.

COMPANIES WITH SUPERIOR PROCESS MANAGEMENT RELY ON SMART IT APPLICATIONS AND HIGH UTILIZATION

How do manufacturing companies with superior process management use IT in their core business processes? Where do they stand out against their competitors?

After three decades of data processing in the manufacturing sector, the technological standard of the IT companies use is high. Today an extensive IT infrastructure is available in all companies surveyed. Apart from conventional mainframes, local PC networks and special-purpose computer networks (e.g., for simulation) have become firmly established systems (see Exhibit II–2). Most companies with sales above $200 million have already switched from mainframes to client/server solutions, a trend that will continue.

Ongoing trend: Migration to client/server system architecture

IT applications in the manufacturing sector have been developed beyond those conventionally used in accounting, sales, and materials and logistics management. IT has played an important role in product development since the early 1980s, when CAD systems were introduced. In production, IT applications have reached the level of shop floor control.

The migration from mainframes to client/server architecture is being driven by the overwhelming trend toward standardization and integration of user software. Today 66 percent of the companies surveyed use integrated standard software as their main system (see Exhibit II–3). They use, for example, SAP R/3, Baan IV, Oracle Applications, SSA BPCS, J.D. Edwards, and JBA. Standard software is widely

Most companies use standard software.

Exhibit II–2
Type of Hardware Used

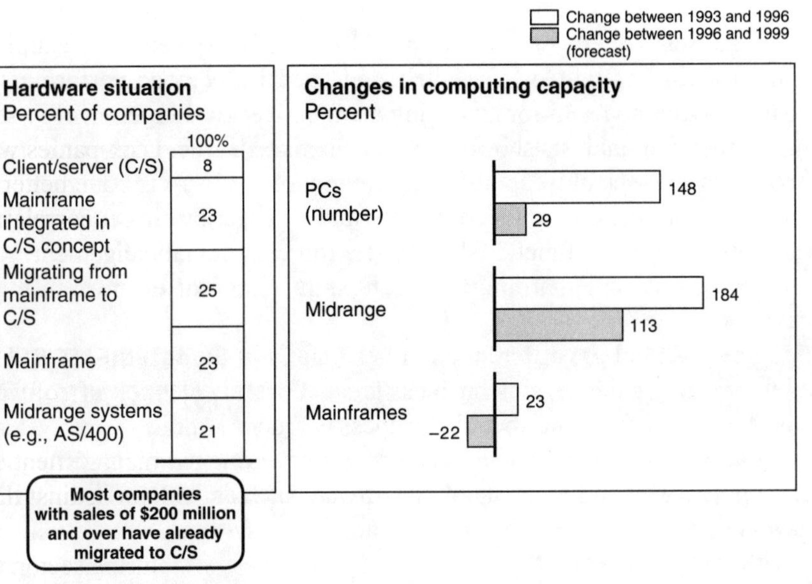

Source: McKinsey & Company, Inc.

Exhibit II–3
Type of Software Used

Percent of companies surveyed

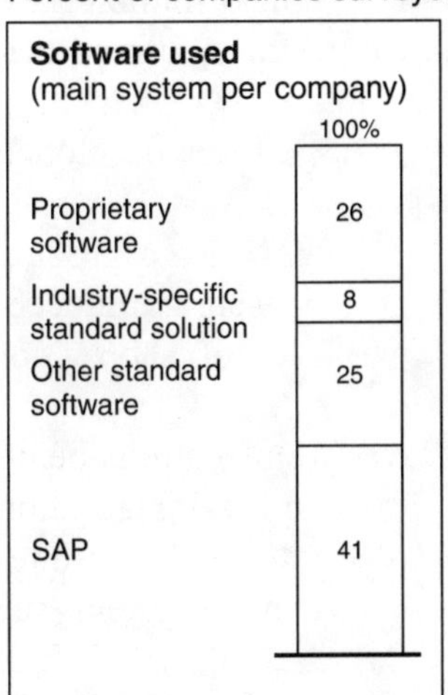

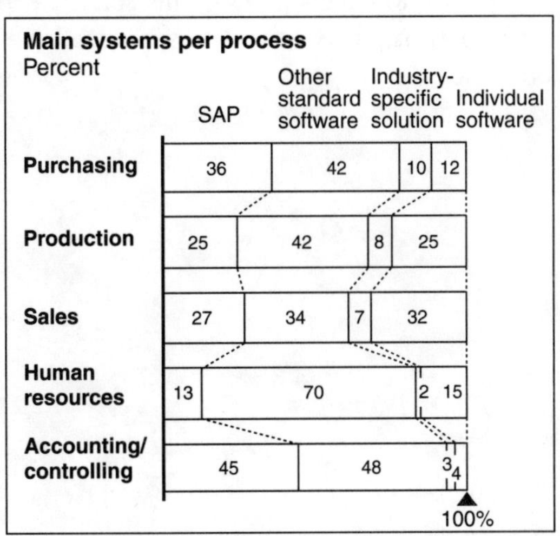

Source: McKinsey & Company, Inc.

used in all business processes, especially in administrative functions such as accounting and human resources management, where the figure is 80 percent.

This might seem to suggest that IT has become a kind of standard infrastructure, bought off the shelf and used by companies across the board, leaving no room for achieving competitive advantage.

In fact, our analysis showed that the opposite is true. Used in the right way, IT is a powerful weapon for differentiation—vis-à-vis both competitors and customers. IT stars constantly find ways to introduce IT applications with superior functionality faster, more systematically, and more efficiently, thus safeguarding their competitive edge or even capturing a new advantage.

The potential for differentiation lies mainly in the functionality of IT applications for the various business areas. Careful selection of software based on benefits for the specific business is crucial. Additionally, systems integration of different functional areas is a key issue with high differentiation potential. Vendors of standard software such as SAP or Baan, which have largely built their business concept on systems integration, are greatly benefiting from this.

However, the selection of IT application systems alone is not sufficient to achieve real differentiation. In addition to IT functionality and availability, systems utilization is of great importance. The best of the best differentiate themselves significantly from other IT stars by training and involving users in systems selection and implementation, as we shall see in Rule 7.

The following chapters in this section explain the four rules for superior IT performance that relate specifically to a manufacturing company's core processes.

RULE 1

MAKE IT A PRIORITY IN PRODUCT DEVELOPMENT

Many companies in the manufacturing industry still view IT in product development largely as a tool to support isolated tasks and improve efficiency. But the world has changed. The fate of a new product can depend purely on time to market, making effective use of simulation and calculation software and product data management vital. Technological excellence may build crucially on the cross-fertilization between partners that can be achieved only by state-of-the-art communication and coordination. IT has become a strategic weapon in the product development process, and best-practice companies exemplify how important it is to boost the effectiveness of IT in this core process.

IT AS THE KEY DRIVER OF SUCCESSFUL PRODUCT DEVELOPMENT

It is not in itself surprising that successful product developers achieve better financial results than companies with less successful product development. But what empirical evidence backs up this qualitative correlation? To establish this, we used a process indicator of product development performance (see Exhibit B–2) that covers common indicators such as development cycle time, qual-

Successful developers achieve better financial results

Exhibit II–4
Correlation Between Product Development Performance and Corporate Success

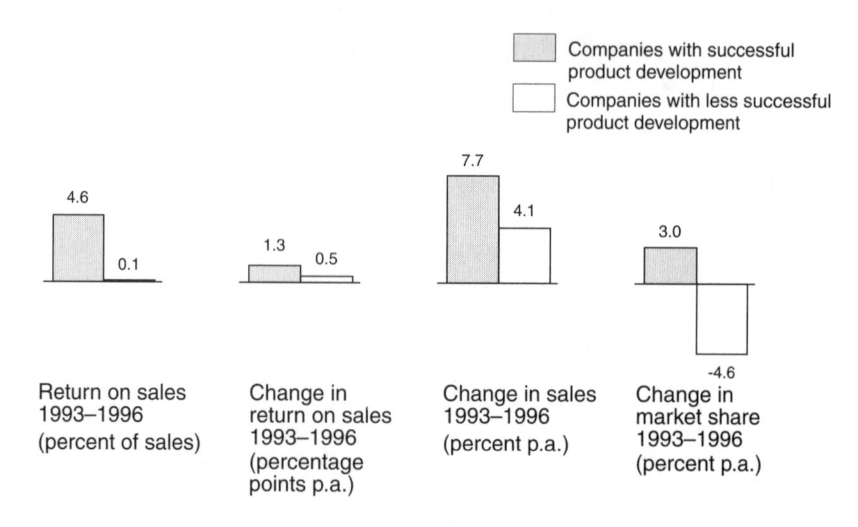

Source: McKinsey & Company, Inc.

ity, and costs of product development. We then compared this with the financial success indicator consisting of the four criteria already described. The findings are impressive (see Exhibit II–4). Between 1993 and 1996 the companies with superior product development performance achieved on average a return on sales of 4.6 percent and annual sales growth of 7.7 percent. Less successful product developers merely broke even, achieving a disappointing profitability of 0.1 percent, with annual sales growth of 4.1 percent.

On the face of it, 4.1 percent annual sales growth does not sound bad. But a comparison of company sales with development in the relevant markets showed that successful product developers increased their market share substantially, while less successful product developers were constantly losing market share.

To determine the relationship between product development performance and IT performance, we analyzed the four IT cultures. Our research

Exhibit II–5
Product Development Performance of the Four IT Cultures

Companies with high IT effectiveness are more likely to be successful developers

Percent of companies surveyed

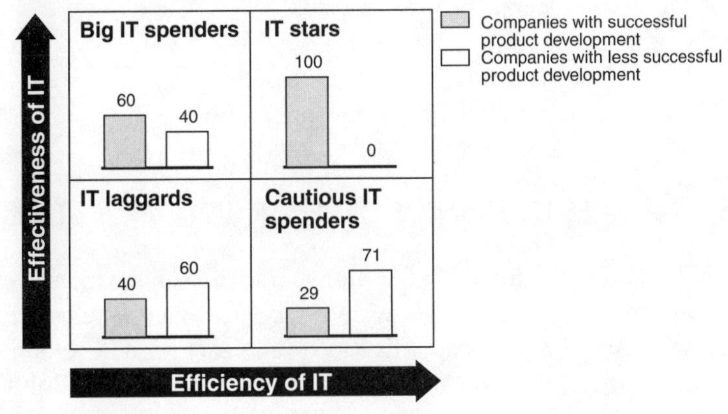

Source: McKinsey & Company, Inc.

IT effectiveness is clearly correlated to product development performance

shows that *all* IT stars are superior performers in product development, while 60 percent of IT laggards are less successful product developers (see Exhibit II–5). A comparison between big and cautious IT spenders proves that *only* high IT effectiveness is positively correlated with successful product development performance. The conclusion is clear: more effective use of IT is the key driver of superior product development performance.

CHANGING ROLE OF IT: FROM ROUTINE OPERATION TO STRATEGIC TOOL FOR PRODUCT DEVELOPMENT

Not so long ago, the benefits of IT in product development could be summarized as faster blueprints, greater design precision, and simpler modifications. As long as IT was used primarily as a tool for improving productivity, this simple formula worked. IT has lost its "nice-to-have" status and has become an indispensable tool in the day-to-day business of product development. In fact, the success or failure of development projects now depends heavily on the scope and type of IT support. Product development strategies that are not based on a systematic concept for the use of IT are bound to fail. We believe selecting the right IT support in this field is becoming key to survival due to the following five factors:

- Reduction of development cycle time
- Merging of technologies
- Optimization of product quality
- Design for manufacturability and assembly
- Growing importance of systems integration

REDUCTION OF DEVELOPMENT CYCLE TIME

Confronted with global competition, companies in the manufacturing sector are forced to reduce the cycle time of product development dramatically. Until a few years ago, five to seven years for a development project in the automobile industry was perfectly common. Now this would be unacceptable. To meet market requirements, companies have to focus on concurrent engineering (CE): performing design and product development activities simultaneously—activities that were previously carried out sequentially. As a consequence, the need for coordination and communication has soared, both within the product development department and between all the other functions involved.

MERGING OF TECHNOLOGIES

Today many products require the integration of very different technologies. A camera, until recently purely an optomechanical product, is today a high-tech camcorder that combines electronics, software, optics, and mechanics in miniature. These demands usually exceed the technical competence of a single company. The answer is often an alliance of various manufacturers, all contributing their specific technological competencies. But this in turn requires intense intercompany communication and coordination.

OPTIMIZATION OF PRODUCT QUALITY

Frequently the optimization of product quality involves conflicting objectives. A car is supposed to be safe, comfortable, quiet, and economical. But you cannot meet the safety requirements by simply using thicker sheet steel:

this will increase fuel consumption. Sophisticated technology design solutions are needed against the backdrop of increasingly complex requirements in a global market. To do this, powerful calculation tools are necessary for simulating and predicting the interaction of multiple features.

DESIGN FOR MANUFACTURABILITY AND ASSEMBLY

The best technical solution is useless if it is too expensive and remains on the shelf. Eighty percent of product costs are determined in the product development process. By specifying the product design, a product developer substantially influences product complexity, materials input, and the manufacturability of a product. The concept of design for manufacturability and assembly (DFMA) offers vast potential for cutting costs. Additional cost savings can be achieved by avoiding problems in the start-up phase; the target must be fast error detection and correction in the design phase and preseries, ideally eradicating errors in serial operation. Vital IT systems here are parts databases, systematic product data management (PDM) for product documentation, and IT communication tools that optimize information exchange among all the parties involved.

GROWING IMPORTANCE OF SYSTEMS INTEGRATION

To avoid being downgraded to a second-tier supplier, suppliers in key industries have to develop competence as a systems integrator. Automotive suppliers that in the past delivered only the interior fittings of doors today find themselves asked to take on responsibility for complete door systems. This change in roles can only be achieved through cooperation with subcontractors and component manufacturers. These product development activities must be coordinated via IT and integrated into the development process of the systems integrator.

MORE EFFECTIVE USE OF IT IN PRODUCT DEVELOPMENT

From a management perspective, the product development process is a complex optimization problem. The specified product characteristics, including the add-on features necessary to gain competitive advantage, must be realized as quickly as possible while keeping to strict cost targets. This requires the systematic cooperation of many different parties: suppliers, purchasing, manufacturing, marketing, sales, and, of course, customers.

Coordination and communication problems may ultimately result in the loss of customers, who simply buy somewhere else. Faced with increasing competition between companies with comparable product development competence, more effective use of IT is crucial for successful product development. The right IT tools allow optimization of product design, ensure the exchange of information between the parties involved, and enable coordination of operations.

USE OF SIMULATION TOOLS

We found a rule of thumb that applies to all phases of the product development process: Successful product developers use simulation tools more frequently and earlier than less successful product developers. Let us look at the use of simulation tools in the different phases of product development.

Successful product developers use simulation tools more frequently and earlier than their competitors do

CONCEPT PHASE

Ninety-five percent of successful product developers that use technical calculations and simulation tools use these in the predevelopment phase, when the basic characteristics of a new product are determined. This phase focuses on increasing the perceived customer value and differentiation from competitors. This means dealing with key issues regarding the product concept rather than specifying details. Using IT to simulate potential product characteristics on screen under the constraints defined by customers or the marketing department can provide the leading edge.

This example is a type of "branching-and-bounding" approach, an instrument used in operations research. The simulations described in the box above go far beyond normal applications in the concept phase, evaluating product features from the perspective of customers and suppliers, and proactively avoiding the loops of detailing product concepts that later turn out to be unacceptable.

We found sophisticated IT applications of this type in an early phase of the product development process at only a few companies, but we are

COMPETITIVE EDGE THROUGH SIMULATION

TIRE MANUFACTURER

A tire manufacturer substantially increased its market share in the OEM business by using calculation programs developed in-house. Its proprietary software enables it to reliably forecast and minimize both the noise emission of tires and the fuel consumption of different vehicles even in the early phases of product development. The approach is based on simulation techniques used to evaluate alternative product scenarios, taking into account variables like vehicle weight, suspension, and type of tire profile.

convinced that this will increasingly become an important factor for differentiation.

DESIGN AND DETAILING PHASE

Once a decision on the product concept has been made, the development engineers' design and detailing phase begins. The different product components must be designed and technically specified according to how they will be used.

This phase is the main area of application for simulation tools for visualizing and evaluating product and process characteristics. These include simulation methods such as kinematic or thermodynamic calculations and assembly and manufacturing simulations. About half of the companies we surveyed use simulation techniques of various kinds. Many tools such as FEM calculations and kinematic simulations are already standard, used primarily to ensure competitiveness.

Half the companies surveyed use simulation tools. Standard software should be the No. 1 choice

When deciding on the system to use, our research indicates that standard software should be the first choice because the benefits of proprietary software developed in-house are notoriously overrated. Usually standard software is not only cheaper; it can often deliver much higher performance than proprietary solutions.

Software vendors have made great strides in the development of standard simulation tools. Nevertheless, finding leading-edge solutions is not

VISUALIZING THE FUNCTIONS OF A PACKAGING MACHINE

MECHANICAL ENGINEERING COMPANY

A manufacturer of packaging machines successfully uses functional simulation. This tool allows three-dimensional visualization of the operation of a packaging machine under development. This tool can visualize the interaction between the packaging machine and the packaging material so the machine's functions and product characteristics can be optimized. The simulation tool can also be used to demonstrate the packaging machine to the customer in a configuration that meets his or her requirements. From the on-screen visualization, the customer can directly assess the benefits of the new machine—for example, to increase productivity or lower maintenance costs.

primarily a question of selecting the right software but of stimulating development engineers' creativity to solve complex problems with this standard software by choosing the right set of parameters.

PROTOTYPING/DEVELOPING PRESERIES

Prototypes are used to align the functionality of individual product components and perform feasibility tests for subsequent production steps. For several years IT systems have increasingly been used in this field.

In model construction, rapid prototyping can significantly reduce development costs. A manufacturer of automotive components succeeded in cutting preseries costs by 64 percent through rapid prototyping, while simultaneously reducing delivery time by 75 percent (see Exhibit II–6).

Although the cost-cutting results are impressive, the main benefit of rapid prototyping is the reduction of throughput time (the time from receipt of an order to delivery). In relation to the costs of subsequent series production, all savings on development costs are marginal, but the reduction of delivery time achieved in the preseries can become the key buying factor for customers.

Production and assembly simulations allow the tuning of product design to production processes in the early phases of product development. Simulations in the form of feasibility studies—to check that a design can actually be manufactured on the existing equipment—minimize the need

Exhibit II–6
Benefits of Rapid Prototyping

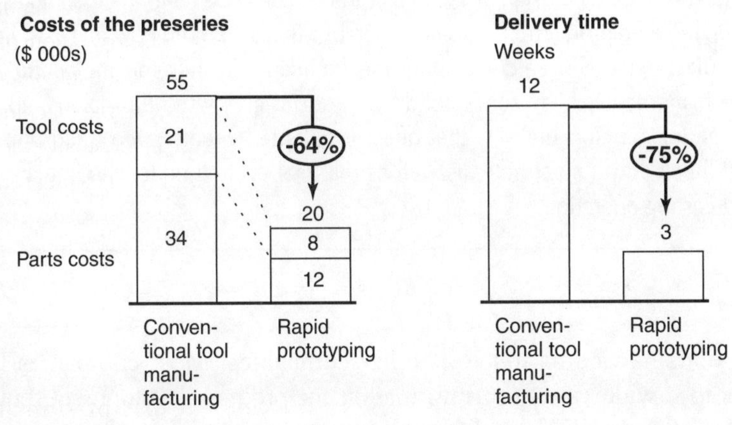

for additional investment and contribute to the development of a robust product design. Optimizing the process layout reduces production costs and improves process stability and quality.

INTEGRATED PRODUCT DATA MANAGEMENT

Contrary to expectations, expert systems and artificial intelligence have not yet been able to break human dominance in the product development process, but the use of IT systems has changed the work of development engineers in the creativity phase. On-line information retrieval, communication, and exchange of ideas with other developers receive considerable support from database systems and communication tools.

CAD integration and product data management (PDM) are the buzzwords in discussions about the integration of IT systems for product development. We found that companies with extensive PDM systems have reduced their development cycle time by 22 percent on average in the last

DESIGN FOR MANUFACTURABILITY

INJECTION-MOLDED PLASTIC PARTS

A manufacturer of injection-molded plastic parts integrated a manufacturing simulation into the design phase of the product development process and as a result reduced production costs considerably. By using mold flow analysis, the manufacturer could calculate how the liquid plastic fills the mold and at which moment which zones solidify. These insights were translated into a design for manufacturability that offered multiple benefits: even mold filling and thus a lower scrap rate, as well as minimal cycle time for the injection-and-casting operation.

three years—twice as much as other companies. Most companies have access to a wide variety of information for product development, but the data are often handled by different systems and are difficult to combine consistently.

This is why integrated product data management is growing in importance. No longer is it sufficient to make and manage blueprints; the real challenge is to ensure complete documentation of all relevant product data. PDM starts in the early development phase with the systematic documentation of development data. It combines data from product development and other functional areas. PDM does not end at product launch; it also incorporates data from later stages of the product life cycle, such as manufacturing, sales, and customer service.

Companies with extensive PDM systems have reduced development cycle time by 22 percent in the last three years

We found that successful product developers intend to expand their data basis, systematically integrating data from downstream departments such as accounting, manufacturing, and quality management. Within just a few years, comprehensive integrated data management, already offered by several PDM systems, could become central to product development (see Exhibit II–7).

Apart from managing engineering data, a PDM system in principle provides interfaces to technical and commercial IT systems. These interfaces permit the full integration of other IT systems with product develop-

Exhibit II–7
Types of Information Provided by Product Data Management Systems

In the future, essential project information will also be available via PDM

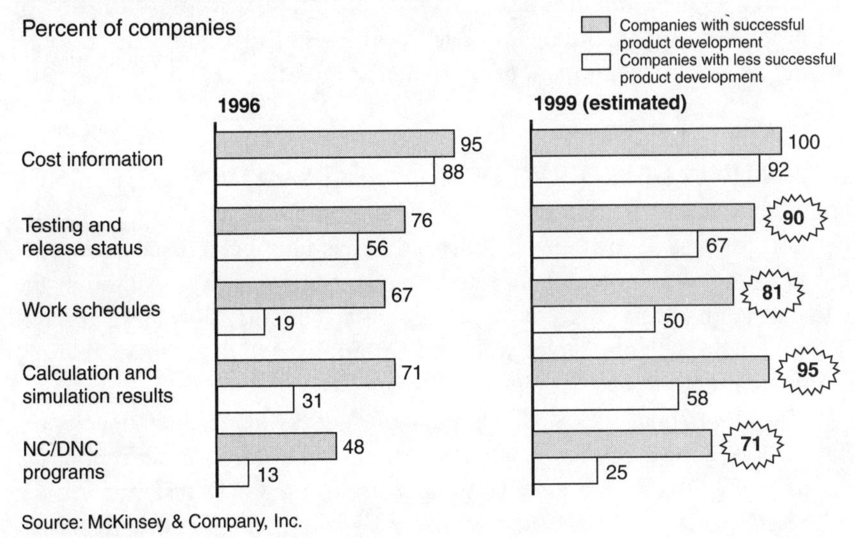

Source: McKinsey & Company, Inc.

ment and support the interaction of these systems and, if necessary, external product development partners' systems.

SYSTEMS INTEGRATION IN PRODUCT DEVELOPMENT

The integration of CAD with engineering databases and simulation tools is important but can pose difficulties. In particular, integrating three-dimensional CAD mechanical designs with two-dimensional electronic layouts is still problematic. But in the face of growing technology integration, integrating different CAD systems is becoming an increasingly urgent task—perhaps the most important reason for standardizing interfaces, a trend that has recently found more and more acceptance.

Simulation functionalities have rarely been integrated into CAD systems. Nevertheless, by linking CAD systems with simulation systems, successful product developers reduce the considerable costs of data preparation required for simulation runs. For instance, 64 percent of successful product developers that carry out kinematic calculations have developed an interface between this system and their CAD system. Successful product developers intend to continue the integration of their CAD systems with other IT systems. However, these interfaces are usually only one-way information highways. The simulation results still must be fed into the CAD system manually. Closing this gap by means of IT, and thus fully integrating the CAD system and simulation tools, is often too laborious an undertaking. Were direct integration possible, an iterative optimization of the product design would be possible, resulting in a quantum leap for optimizing product development cycle time and quality.

INTEGRATION OF PRODUCT DEVELOPMENT AND MANUFACTURING

Great benefits can be expected from the integration of CAD systems with manufacturing's IT systems, a subject that has long been discussed under the heading of computer integrated manufacturing (CIM). Although the CIM concept offers many realistic potential benefits, they have not yet been realized. The objectives were too ambitious, and the idea of the fully automated factory was perhaps too naive.

We found that almost 80 percent of successful product developers have a CAD interface for generating numerical control/direct numerical control (NC/DNC) programs. But direct use of the CAD system for work scheduling and generating bills of materials is rare, and in complex situations seldom leads to good results. Frequently, this is due to incorrect or incomplete data.

Systematic information feedback from manufacturing to product development has only made modest advances so far. Although cross-functional development teams usually ensure that in the product launch phase the required information is available, after series start there is generally no systematic feedback whatsoever from manufacturing to product development. Quality and machine data are seldom analyzed and edited appropriately. Information required by product development for ensuring the manufacturability of the product may not be available, wasting the opportunity to develop robust production processes in an early phase of product development.

The product development process can be better aligned to the production process using PDM information from manufacturing: for example, work schedules or NC/DNC programs. This also helps to capture efficiency potential in production. Our research reveals that such information is available to companies with successful product development almost three times as frequently as to less successful product developers. Successful product developers plan to add all the information required for optimizing the entire process to their IT systems in product development.

INTEGRATION WITH MARKETING, SALES, AND SERVICE

A good knowledge of the market allows an early focus on promising development projects. Successful product developers therefore ensure direct access to relevant market information, and often have the data documented much more systematically (e.g., in product-market databases).

Successful product developers include after-sales service in their information loop

Companies with successful product development also generate distinctly more project ideas. Unlike less successful product developers, they kill development projects that do not clearly meet given objectives early on, thus freeing product development capacity for more attractive projects. These companies stop around 80 percent of their development projects after less than six months, whereas less successful product developers take two or more years to break off many projects, thus tying up scarce product development resources on projects without a chance of success.

Successful product developers include after-sales service in their information loop. If information from the service department is available in the PDM system, the design can be optimized by evaluating service and error reports. This helps to achieve higher product quality and simplifies the future provision of services. Successful product developers use information from after-sales service in the early phases of product development to design products that are suited to efficient maintenance. Companies with good product information, and thus better knowledge of their product configurations, estimate the improvement potential of their service performance through IT to be three times higher than companies without a cross-functional PDM system (see Exhibit II–8).

INTEGRATION WITH CONTROLLING AND PURCHASING

Access to controlling department databases is vital for product costing, but often this information is not available in PDM systems, and requires time, energy, and money to obtain. The same is true for information from purchasing. If development engineers were better informed about the performance of suppliers, they could take any necessary constraints into account during the development phase.

Usually, only the costs of parts are known; process costs and costs of complexity often cannot be estimated until later in the product development process. The integration of IT systems from product development and accounting can provide the information product development requires

Exhibit II–8
Benefits of Product Data Management Systems to Customer Service

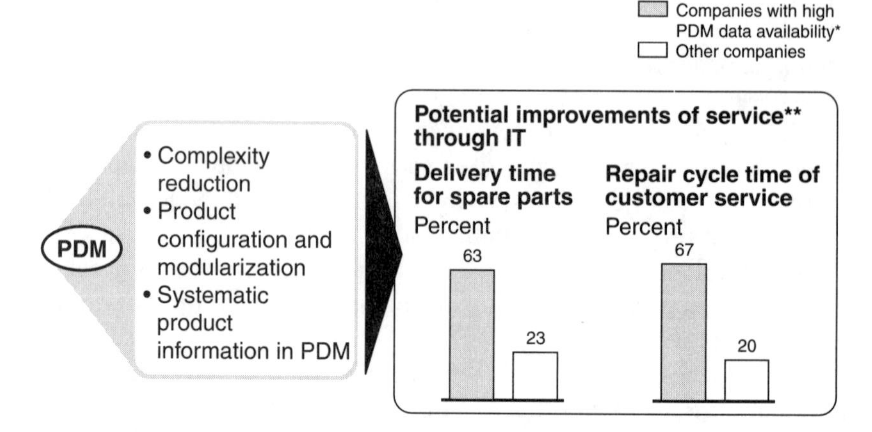

* More than 10 items out of 14 PDM data asked for.
** Assessment of companies surveyed.
Source: McKinsey & Company, Inc.

for an efficient design to cost, and for early IT-based preliminary costing. We found that this can reduce costing risks in the early phases of product development by 50 percent on average.

INTEGRATION WITH DEVELOPMENT PARTNERS

For most companies, product development partnerships are increasingly turning out to be a driver of corporate success. Frequently, in fact, the required systems or technology competence can only be ensured by such partnerships.

Successful product developers have already begun to step up their outsourcing of business functions, even of product development activities. Companies with successful product development incorporate external development partners in cross-functional teams twice as much as less successful product developers (see Exhibit II–9). This is usually well worth the effort. Companies with good communication links to their ex-

> **Successful product developers incorporate external development partners in cross-functional teams twice as frequently**

Exhibit II–9
Outsourcing R&D and Cooperation with Development Partners

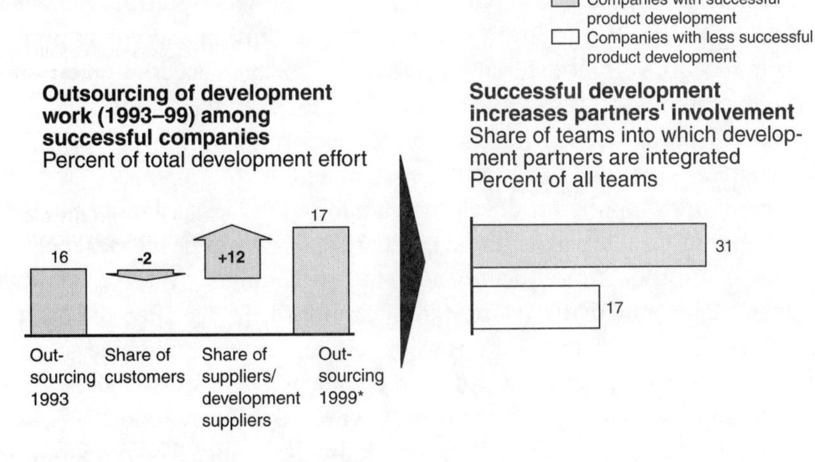

Source: McKinsey & Company, Inc.

ternal development partners have a distinctly higher rate of innovation. Revenues from products developed during the previous twelve months is 20 percent, almost twice as high as at the rest of the companies surveyed.

Nevertheless, this integration happens only under appropriate conditions. Although the demand for exchanging CAD data with product development partners is increasing rapidly, the interface problems between the different CAD systems have not yet been solved. The cost of data exchange, including preprocessing and rework of CAD data, can run up to 30 percent of development costs. This becomes even more significant if you consider that successful product developers intend to double their data exchange within the next three years—after quintupling data exchange in the past three years.

Systems integration of this type embraces all technical and commercial product data from the entire product development process, including external product development partners.

Systems integration can contribute to corporate success in multiple ways. It can increase development effectiveness, avoid late design changes, and increase production efficiency through complexity management, to take just three examples that we will now examine.

INCREASING DEVELOPMENT EFFECTIVENESS

The shared use of PDM systems with external product development partners probably has the greatest effect on development effectiveness—above all, the reduction of development cycle time and improvement of quality. But the risks are considerable, especially in protecting critical development know-how. We discovered that about one-third of successful product developers allow their external development partners access to their product development databases, and others are planning to introduce this aspect of knowledge sharing. Successful performers differ markedly from companies with less successful development in this respect. From the viewpoint of a supplier as development partner, earlier and more intensive involvement on the part of the customer (and system developer) reduces the costs of product modifications in subsequent phases of product development. This especially avoids product modifications after the start of series production. Time and costs for coordination are reduced substantially. Joint product specification also enhances the customer value of the new product, frequently in combination with a significant reduction in production costs. The Boeing 777 is a well-known example. The time for error correction and rework in this development project was reduced by 50 percent, and the improved information contributed vitally to eliminating 12,000 errors early in the development process.

Good availability of information and a close involvement of external product development partners are essential prerequisites for high product quality and stable production processes. Companies with high PDM data

CONCURRENT ENGINEERING THROUGH PDM

MANUFACTURER OF AUTOMOBILE ENGINES

A manufacturer of automobile engines started a product development project with the intention of integrating design, drafting, manufacturing, and customer service using a single workstation-based PDM system, and providing product and engineering data for an entire network of companies. The close IT-based integration of customer(s) and suppliers led to the establishment of some innovative processes. The most outstanding result was that time to market was halved, while product quality was improved significantly.

availability and companies with close IT-based connections to external partners had, on average, a distinctly lower reject and rework rate (see Exhibit II-10). Close and early involvement of external product development partners also leads to much higher design stability. This makes it possible to gear production better to the product and create more stable production processes. This in turn reduces the customer rejection rate when compared to the figures of other companies (see Exhibit II–10).

AVOIDING LATE DESIGN CHANGES

Companies that use IT to improve integration and communication are able to reduce modifications of product specifications in later phases of product development. By using development information better, they can substantially shift modifications to the early phases of the product development process (see Exhibit II–11).

This allows these companies to combine short development cycle times with a high "maturity" of the product at the time of the product launch. The number of modifications of product specifications during detailed planning, the test phase, and the first six months after ramp-up can be cut in half, greatly reducing the costs of product development and production.

Exhibit II–10
Impact of PDM-Based Information on Production Quality

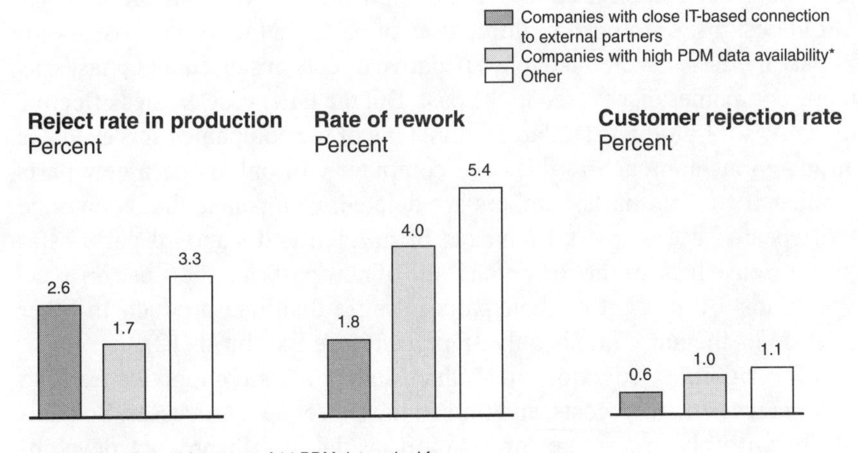

Exhibit II–11
Impact of Integration and Information Availability

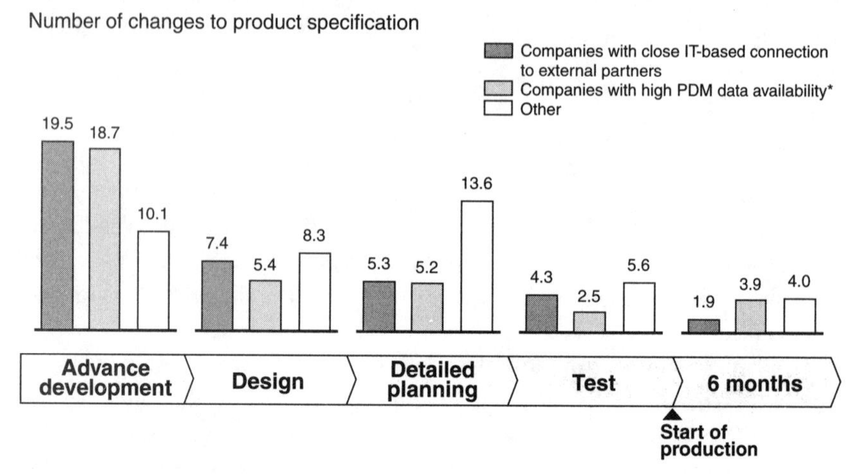

* More than 10 items out of 14 PDM data asked for.
Source: McKinsey & Company, Inc.

INCREASING PRODUCTION EFFICIENCY THROUGH COMPLEXITY MANAGEMENT

Frequently the reason for differences in IT performance lies less in the type of IT support than in the goal pursued by using IT. For example, the same parts database can be used to generate new product versions on the basis of old designs, or to push multiple uses of parts. In the former case, complexity increases, which will have negative effects in subsequent phases, as many companies discovered in the past. But the latter case creates effective incentives for product standardization. One of the companies surveyed has an ingenious approach to this. The computer will only issue a new parts number if two old parts numbers are deleted. Companies that encourage the repeated use of parts by means of normed and standard parts catalogues, reuse lists, or the restricted issue of new parts numbers use on average almost 70 percent of their parts in more than one product. In other companies the reuse rate is only 35 percent (see Exhibit II–12).

The rigorous reduction of product and process complexity leads to advantages in quality, costs, and time in the subsequent processes. So companies with above-average information availability in product development succeed in reducing the share of single-item production in favor of

REDUCING PRODUCT COMPLEXITY

REGULATING AND CONTROLLING EQUIPMENT

A manufacturer of regulating and controlling equipment succeeded in shifting the complexity of all the product variants of one product line to the software embedded in the product. The resulting reduction in parts variants led to substantial cost savings in purchasing, manufacturing, logistics, and service.

Exhibit II–12
The Right IT Support Encourages Reuse of Components

The benefits of engineering data management support

Companies with high degree of IT utilization for reuse of parts*
Companies with low degree of IT utilization for reuse of parts

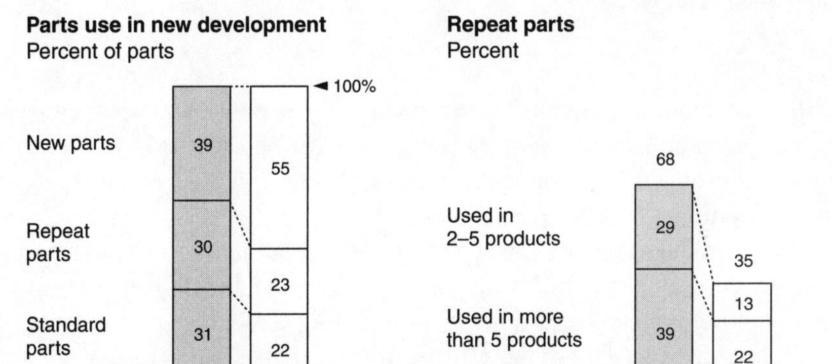

* Use of normed and standard parts catalogues, match searches, repeated use lists, and restricted issue of parts numbers.
Source: McKinsey & Company, Inc.

a greater continuous series-type production and continuous batch production, thus achieving bigger batch sizes in the manufacture of intermediate goods and final assembly. In companies that have information about inspection and release status, activity-based costs, work schedules, NC/DNC programs, as well as calculation and simulation results in PDM systems, we generally observed a much more stable production process.

Exhibit II–13
Impact of IT Support on Manufacturing Lot Size

Availability of cross-functional information increases manufacturing lot sizes

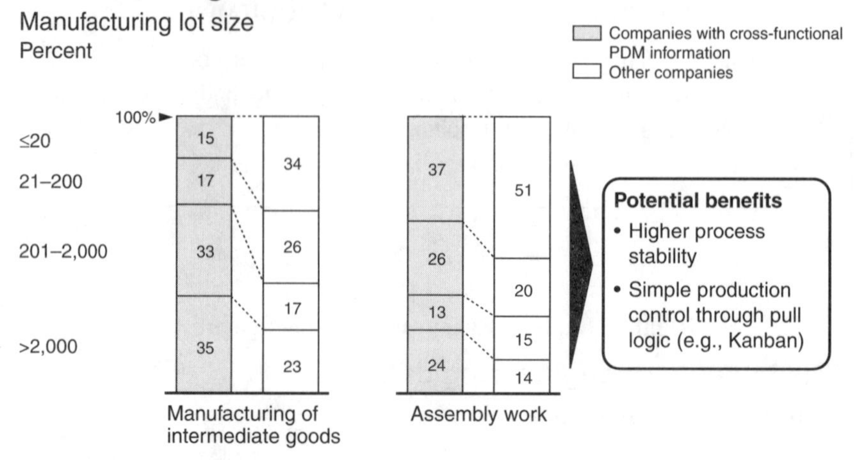

* Cross-functional PDM information: cost data, testing and release status, work schedules, calculation and simulation results, NC/DNC programs.
Source: McKinsey & Company, Inc.

The share of batch sizes in prefabrication below 200 pieces is 32 percent—only half as much as in companies with less cross-functional information available to product development. However, the differences between batch sizes in assembly are lower (see Exhibit II–13).

Better product development, production, and market information enable successful product developers to assess the required variety of products more precisely and represent this through modular product configurations geared to the production facilities available.

IT SUPPORT FOR COORDINATION AND INFORMATION FLOW

Close partnerships in product development, between different business functions, and especially between different companies require cooperation that goes beyond mere exchange of data. Relevant project information must be made available to all parties involved in product development. The communication between product development partners can be greatly improved when IT is used: from e-mail and project planning systems through to groupware systems and shared work flow management in product development. New communication media enable the customer to influence the design and functionality of products early in their development.

Driven by OEMs, the automotive supply industry already uses intensive communication and data interchange in product development. The widespread cooperation among companies in this industry makes the use of IT for coordination and planning particularly useful, but it is still largely limited to the exchange of CAD data and communication via e-mail. And within the companies themselves, product development tends to be isolated from an IT perspective.

Successful product developers are far ahead. They use the whole spectrum of IT to intensively involve external partners (including customers and suppliers) in product development (see Exhibit II–14).

USE OF E-MAIL AND GROUPWARE

We found that 70 percent of companies surveyed use e-mail for communication with their product development partners. However, the functions of groupware systems are seldom used, and usually only internally. E-mail can certainly improve efficiency, allowing swift and paperless transmission of messages, but the use of groupware can lead to completely new processes, thus stimulating improvements in effectiveness.

Exhibit II–14
Joint Use of IT Systems with External Partners

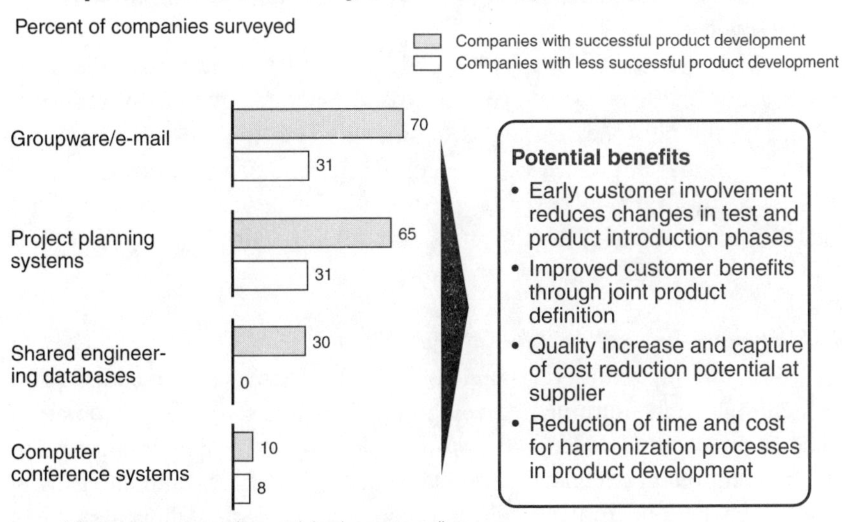

* Customers, parts suppliers, and development suppliers.
Source: McKinsey & Company, Inc.

FAST ERROR ANALYSIS THROUGH GROUPWARE

PLANT ENGINEERING COMPANY

By using a groupware system, a plant engineering company has reduced the time required for error detection substantially. When problems arise, customer service, sales, and other departments feed a report with an error description, including scanned sketches and documents, into the system. First, the information is distributed within the local organization. If nobody can solve the problem, the message is then distributed to other business units. When solutions are found, they are added to the file. In addition to resolving problems quickly, this system ensures that solutions are documented and readily available for use in similar situations and product development in the future.

TRANSITION TO WORK FLOW MANAGEMENT SYSTEMS

The use of work flow management systems in product development is in its early days. In view of the increasing tendency to integrate IT systems in product development, the future benefits of work flow systems for process integration and standardization will be very high. Systematizing the use of work flow systems is likely to yield a further dramatic reduction in development cycle times.

To what extent a company can really benefit from these new possibilities depends on the corporate culture. Work flow management systems can help to optimize processes across companies only if thinking in conventional functional boundaries is overcome. The use of work flow management systems in product development cannot bring about this paradigm shift alone, but it can certainly act as a driver.

SHARED USE OF PROJECT PLANNING TOOLS

We discovered that around 80 percent of all companies surveyed use project planning tools in product development. The decisive difference is how these tools are used. Successful product developers rely on project planning developed together with their external product development partners, driving process optimization across company boundaries. They also try to

CHANGE MANAGEMENT AS AN IT-BASED PROCESS

MECHANICAL ENGINEERING COMPANY

Using a work flow management system, a mechanical engineering manufacturer succeeded in significantly reducing the throughput time for modifications. A standard process for changes and modifications was introduced that is an explicit component of the work flow system. Apart from many modification forms, idle times in the modification process were eliminated. The IT system ensures that only authorized designers can carry out modifications, and prevents modifications after design freeze. In addition, an e-mail function informs all team members automatically about modifications to the master model.

capture synergies between different development projects. Half of all successful performers use network planning techniques or multiproject management, whereas less successful product developers focus on individual projects and neglect potential synergies.

Half of all successful performers use network planning techniques or multiproject management

COMMUNICATION SUPPORT VIA COMPUTER AND VIDEO CONFERENCES

As a whole, the companies surveyed make little use of computer and video conferences; personal contact between the development engineers appears to be much more important. But computer and video conferencing can speed up coordination processes tremendously: teams can hold virtual meetings as often as they want.

In the future, computer conferences will open up vast opportunities when it becomes possible not only to use the same graphical user interface but to work jointly on the same data model. Joint design on screen could provide even greater benefits than can be accessed by simply gathering everyone involved at the same physical location.

DATA MINING VIA NETWORKS

All companies surveyed use the Internet in product development, and most have established an intranet. Amazingly, successful product developers are much more restrictive regarding Internet access than less successful product developers (see Exhibit II–15). High performers in product development frequently use specialists who both process information requests and proactively distribute relevant information to different target groups. This approach saves time, produces better search results, avoids redundant search questions, and reduces communication costs.

The attitudes of companies in Europe differ from those in the United States. American companies allow employees broader access to the Internet; the number of R&D staff with access to the Internet is twice as high as at European companies. Perhaps this simply reflects the more advanced, more professional attitude of Americans toward surfing in data networks. In that case we could expect a similar development over time in Europe.

Exhibit II–15
Information Search in R&D via the Internet

Successful developers maintain a small staff of dedicated Internet experts

Companies with successful product development
Companies with less successful product development

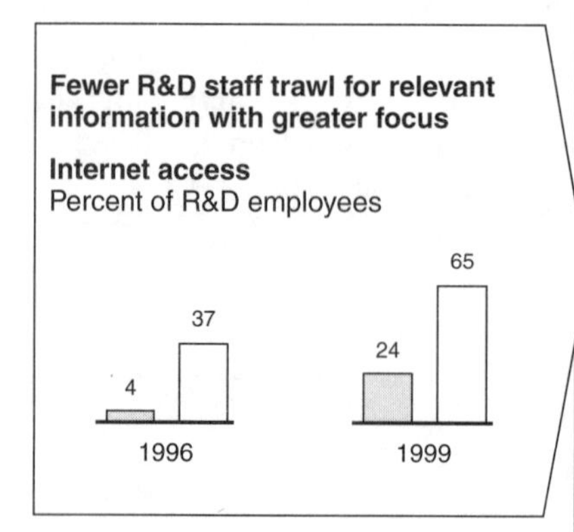

Example:
Two dedicated employees search the Internet for relevant information, work on requests from the functional areas, and place the results in the company intranet

Benefits:
- Time savings and improved information quality by systematizing and focusing the process
- Avoidance of redundant search operations through centralization
- Reduction of communication and hardware costs
- Increased security by decoupling intranet and Internet

Source: McKinsey & Company, Inc.

SUMMARY

The companies in our survey all use IT to improve the efficiency of individual product development activities, with varying impact. Successful product developers stand out from the rest because they understand the necessity of systematic, integrated IT support. They have succeeded in integrating the IT systems of different departments (e.g., product development, manufacturing, marketing) and ensuring IT-assisted coordination beyond company boundaries. Best-practice examples illustrate the potential benefits that can be tapped through internal and external integration of product development activities and systematic use of new IT tools (see Exhibit II–16).

The trend toward integration of IT systems is especially focusing on PDM systems. Other IT applications, such as project management tools, work flow management systems, and communications support, will also be integrated gradually. The vendors of CAD systems, simulation tools, and commercial IT systems are increasingly taking into account these integration trends and incorporating PDM functionalities in their systems. As a result, product data management is becoming an integrated effort of even more different applications and vendors building a truly product-specific knowledge database.

Exhibit II–16
IT As a Strategic Tool for Product Development: Summary

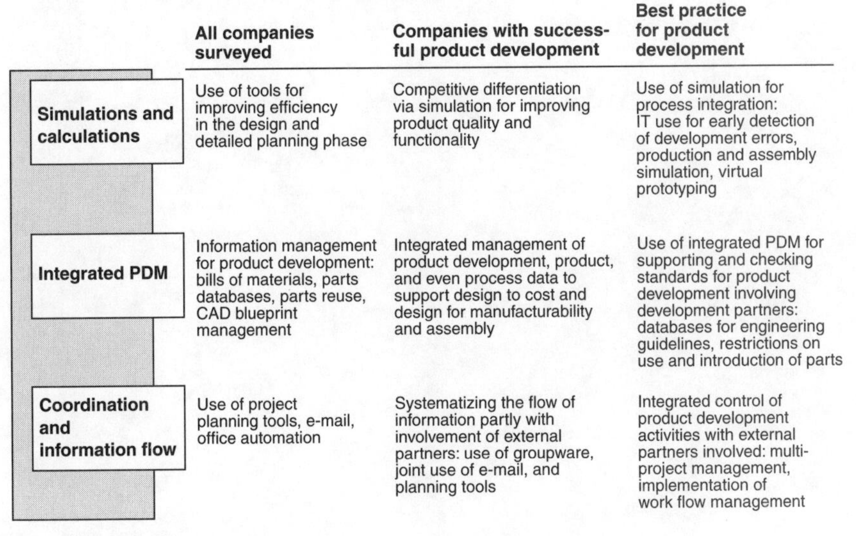

IT has emerged as a strategic tool

	All companies surveyed	Companies with successful product development	Best practice for product development
Simulations and calculations	Use of tools for improving efficiency in the design and detailed planning phase	Competitive differentiation via simulation for improving product quality and functionality	Use of simulation for process integration: IT use for early detection of development errors, production and assembly simulation, virtual prototyping
Integrated PDM	Information management for product development: bills of materials, parts databases, parts reuse, CAD blueprint management	Integrated management of product development, product, and even process data to support design to cost and design for manufacturability and assembly	Use of integrated PDM for supporting and checking standards for product development involving development partners: databases for engineering guidelines, restrictions on use and introduction of parts
Coordination and information flow	Use of project planning tools, e-mail, office automation	Systematizing the flow of information partly with involvement of external partners: use of groupware, joint use of e-mail, and planning tools	Integrated control of product development activities with external partners involved: multi-project management, implementation of work flow management

Source: McKinsey & Company, Inc.

However, managers frequently forget that using state-of-the-art IT alone cannot guarantee an optimal product development process. To capture the huge potential of IT for gains in effectiveness, it is essential to first redesign the product development process, an issue that we will discuss in detail in Rule 7.

RULE 2

INTEGRATE IT INTO MARKETING, SALES, AND SERVICE

Marketing, sales, and service are traditionally an area of low IT penetration in the manufacturing sector. Nevertheless, the situation here is also characterized by enormous performance gaps between successful and less successful performers. Successful performers achieve revenues per sales representative that are three times as high as those of less successful performers, and invest 50 percent more time in dialogue and cooperation with their customers. Measured against revenues, their marketing and sales costs are 30 percent lower than those of performers that bring up the rear.

Significant IT applications—if they are used at all—are often installed only to improve efficiency—for example, in order intake or on field staff laptops. But often it is difficult to identify promising IT applications if IT is looked at only from an efficiency perspective because the levers for improving effectiveness, the more interesting quest, will remain unrecognized. There is great potential to leverage the capabilities of IT in marketing and sales if companies move beyond simply administering customer data and begin to use their databases for direct marketing or to analyze the sales potential of customers. However, the data management this requires cannot be performed without high-performance IT support.

Our findings for after-sales service were similar. What counts is not so much efficiency as effectiveness: functionality of the IT systems used in the field, and how usefully the data have been consolidated from other relevant functions. Utilization at companies with successful service performance is twice as high as at less successful performers. The trend toward high-tech applications at stars is also clear, such as on-line error diagnosis or embedded software for repair and maintenance, thus greatly enhancing customer convenience.

IT SUPPORT FOR MARKETING AND SALES

The biggest barrier to more effective IT use in marketing and sales is often the lack of data integration. In principle, a lot of the required data should be available in a company, but they are dispersed over many IT applications—in the customer database of the field sales force, the accounts receivable data of the accounting department, or the vendor card file of the procurement department. And these data are in a multitude of different formats. If you need sales statistics or a profit report broken down by product, region, or customer, the undertaking is like trying to make a picture from the muddled pieces of different jigsaw puzzles: some pieces are missing, some are duplicates, and some do not seem to fit.

45 percent of successful performers in marketing and sales rely on off-the-shelf solutions, compared to 25 percent of less successful companies

In order to escape this dilemma, many companies use integrated standard software. About 45 percent of successful performers in marketing and sales rely on off-the-shelf solutions, in comparison to 25 percent of less successful performers. When data integration through integrated standard software reaches its limit, successful performers in marketing and sales employ data warehousing provided they can cope with the technical requirements. This approach combines data from different applications in one relational database and allows processing of data for new applications or analyses.

This form of data consolidation and integration has recently become economically feasible due to the plunge in prices for storage media and computing power. Successful performers in marketing and sales generally use the consolidated data from the data warehouse for reporting and analyzing customer behavior—for example, for identifying negative patterns in the buying behavior of customers. Instead of creating a single data source, many companies use data warehousing merely to consolidate their different databases that could not be processed in their previous form.

If data warehousing is used only as an interface between different legacy applications, its potential is not fully exploited. Data warehousing could become one of the most important tools for customer relationship management (CRM). The CRM approach, originally developed in the consumer goods industry, is increasingly penetrating other industries. CRM focuses on the micro segmentation of customers down to the segment of one—the individual customer. The intention is to develop more targeted, customer-oriented marketing to approach these segments much more effectively. Applying the CRM approach requires an intensive use of IT, which only a few companies in the manufacturing sector have been able to afford to date. Data warehousing is only the data platform from which companies can search for the proverbial needle in a haystack using multilevel filters and sophisticated statistical techniques.

CANVASSING CUSTOMERS

Many of the companies surveyed do not draw on the full potential of IT to support sales canvassing. Consequently, they regularly lack crucial information because sales staff do not have access to all marketing and sales data simultaneously. More than 50 percent of successful performers in marketing and sales do not use an integrated sales information system (SIS); instead, they use at least two IT application systems. About 13 percent of the companies surveyed use four or more different IT application systems in marketing and sales—clearly a severe handicap.

Although 60 to 70 percent of all sales representatives now have laptops and mobile phones, they are often no better informed than in the past. Generally only basic data (e.g., prices and customer master data) are available on the computer—data that in the past were distributed on paper. Details on product specifications and variants are rarely available. Information that can prove decisive, such as sales leads from inquiries by phone, responses from direct marketing activities, or analyses of customer potential, is usually unavailable.

SUCCESSFUL PERFORMERS USE IT TO BUILD CLOSER CUSTOMER RELATIONSHIPS

Successful performers in marketing and sales take full advantage of processed and customized data. Their mobile offices are only equipped with information that responds directly to the needs of customers: "Which products are available for immediate delivery?" or "What production capacity is available at present?" or "How long is the delivery time for make-to-order production?" Having the answers to these questions accessible at the touch of a key simplifies the dialogue with the potential buyer and offers sales representatives what they need most: more time for the customer.

There are big differences in how sales representatives work with their customers. On average, sales representatives of successful performers in marketing and sales spend 46 percent of their time with customers, while the figure for less successful performers is only 28 percent. Sixty-seven percent of the successful performers use IT-assisted production scheduling, compared to 44 percent at less successful performers. Additionally, 46 percent of successful performers use IT-assisted feasibility checks, in contrast to 16 percent of less successful performers (see Exhibit II–17). Instead of having to discuss order data and specifications with customers, successful performers can use their time to develop relationships and foster customer loyalty.

MORE EFFECTIVE USE OF IT WILL FURTHER ENHANCE CUSTOMER ORIENTATION

Exciting new IT applications are emerging that will revolutionize the customer interface in this industry (as in many others). Let us look more closely at two of them.

PRODUCT CONFIGURATORS

A product configurator allows sales reps to present an entire product line on screen when they visit the customer—their full range of hydraulic cylinders, for example, or all the variations on passenger bus equipment available. On the spot, the sales rep can work out costs and delivery dates for an offer tailored to the customer's wishes.

The value to the customer is obvious. Offers can be prepared faster and better, and unpleasant surprises can be avoided—for example, finding out belatedly that the required configuration cannot be delivered, or can be manufactured only at huge extra costs. If the product configurator is supplemented by appropriate costing data, an exact price can be agreed upon with the customer there and then.

The smart use of product configurators can boost sales performance. For example, after implementating a product configurator, a manufacturer of buses achieved a decline in the number of special customer requests by 90 percent and reduced the time spent clarifying orders by 50 percent. One of the reasons is that preparing the configura-

Using a product configurator reduced the number of special customer requests by 90 percent

Exhibit II–17
Use of Integrated Sales Information Systems (SIS)

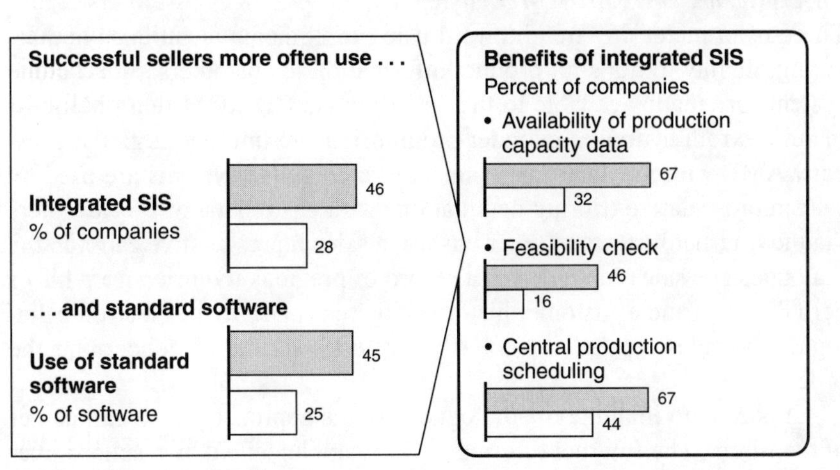

Source: McKinsey & Company, Inc.

tor guidelines often involves streamlining the product architecture and eliminating exotic offerings, which helps to reduce product complexity and manufacturing costs. As well as compensating for deficits in technical know-how, product configurators also help to counteract the personal preferences and (sometimes unoriginal) routines of sales representatives.

The benefits of this tool are so convincing that some companies have gone one step further and made the product configurator available to customers. Cisco, a leading U.S. manufacturer of network components, has integrated a product configurator into its Internet ordering systems that sells complex products costing up to $100,000. This configuration supports customers in making their selection and checking orders. In 1997 Cisco achieved sales revenues of some US $3 billion via this order system. This example demonstrates the potential that exists for other technical capital goods—for example, in mechanical engineering, and for complex consumer goods, such as cars or kitchen installations.

Product configurators are not yet widely used. One reason is that you first have to define your company-specific product range and feed it into a computer, a task that cannot be performed by standard software alone. But the costs of this software development are justified by the wide range of applications. Product configurators not only are suitable for supporting sales representatives and serving as an Internet application for direct order intake via digital sales channels but can also be used as a component of the ERP system to check all incoming orders.

MULTIMEDIA MARKETING TECHNIQUES

There is an increasing trend toward using multimedia techniques in marketing, mainly to transmit product information to customers. Stand-alone systems are made available to the customer via CD-ROM (e.g., Mannesmann Rexroth hydraulic cylinder program) or the Internet (e.g., the company AMP with its electronic connectors). Dialogue systems are used by sales representatives during negotiations with customers, or by customers via the Internet. If the customer wishes, a sales representative can join the dialogue. The sales rep receives a record of previous inquiries from his or her IT system and a customer history (if there is one). In the electronic dialogue, the sales rep helps to clarify any open questions concerning the order.

It is easy to imagine the dialogue systems approach as the call center of the future. The Internet offering will be supplemented by a seller-buyer dialogue (e.g., using a picture phone via the Internet). Although the required hardware and software exist in pilot versions, actual applications will probably not be available on a large scale for several years.

ELECTRONIC DATA INTERCHANGE ORDER TRANSMISSION

In most of the companies surveyed, order bookings and subsequent production scheduling are done manually using fax and mail support, with the inevitable disadvantages of duplicated work, input errors, and longer throughput time. Electronic data interchange (EDI) is a proven technology for the transmission of orders. Less successful performers in sales appear to still view this technology with some skepticism.

Successful performers in order processing understand that it is not the number of customers with an EDI connection that matters, but how many transactions are actually processed via EDI. They therefore focus on core customers, who place many orders and have a high share of revenues. The share of orders transmitted via EDI is 72 percent at successful standard product manufacturers and 47 percent at less successful standardizers (see Exhibit II–18).

There is every indication that order processing can be standardized and rationalized using EDI. Among manufacturers of standard products

Exhibit II–18
Use of EDI by Customers for Placing Orders

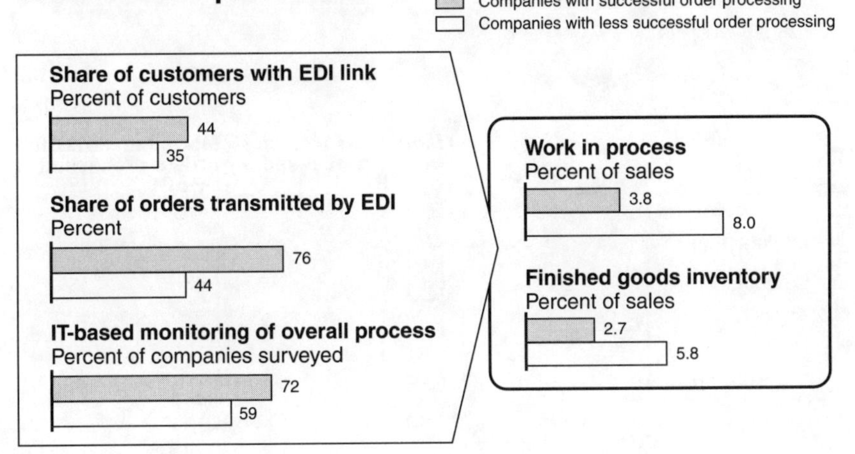

Source: McKinsey & Company, Inc.

that process more than 50 percent of their orders via EDI, the cost of back office sales work is 0.7 percent of revenues (see Exhibit II–19). At comparable companies with an EDI share of only 10 to 50 percent, the costs are 1.4 percent—twice as high.

For hybrid companies, which manufacture both standard products and tailor-made products, the use of EDI offers advantages too. Without EDI, the costs of back office sales work are 4.8 percent of revenues, and with EDI only 2.7 percent.

EDI is no cure-all for inefficient order processing, but is a strong lever for tapping unused potential. The high investment for EDI/EDIFACT (electronic data interchange for administration, commerce, and transport) that is frequently pointed to is due to programming the required interfaces with relevant existing IT applications. But in the long run, the high expenditure seems justified by the better customer connectivity that results.

For those who find this solution too expensive, the Internet will probably offer a more favorable alternative soon; initial applications of Internet EDI have already been launched. Users will benefit from much lower installation and running costs, an important aspect, especially for small- and mid-sized companies.

Exhibit II–19
Cost of Back Office Sales Work in Relation to Use of EDI

Cost of back office sales work can be reduced by use of EDI in order processing

Percent of sales

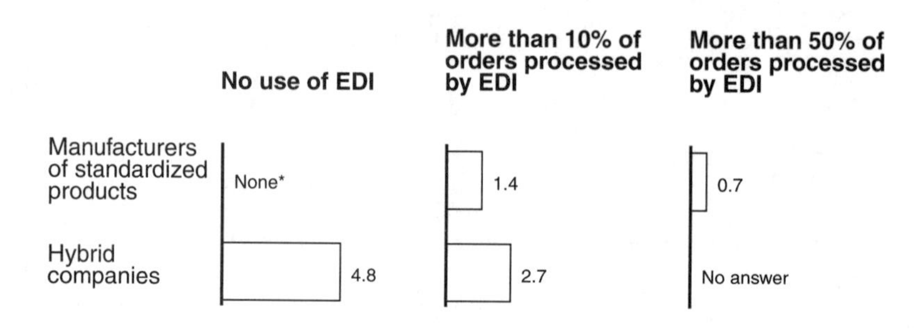

*No participants fall in this category.

Source: McKinsey & Company, Inc.

IT SUPPORT FOR CALL CENTERS

DELL COMPUTER

Michael Dell began manufacturing and selling PCs in 1984. He decided on a business strategy totally different from that of competitors: no sales through dealers, no field sales force for direct sales. Customers could buy Dell's PCs only by phone or fax. This was the crucial point that would determine success or failure.

The call centers of this company had to be extremely efficient in terms of both the quality of the telesales agents and the power of the IT systems; during the phone call, all relevant information about the customer order had to be gathered and checked. In 1994 Dell also installed a purchase order processing system on the Internet. Here, too, the company set new standards for user friendliness, order status information, and delivery time.

In the medium term the embedding of Internet EDI in IT applications for private customers will become commonplace. This system could become the platform for order processing on the World Wide Web, with Internet EDI the standard protocol.

IT SUPPORT FOR AFTER-SALES SERVICE

In the last few years, after-sales service in the manufacturing sector has evolved into a business function with considerable impact on corporate success. A key reason for this change has been rapid margin decline in the product business. As a result, the improvement of customer satisfaction and customer loyalty through excellent service has become a strategic differentiation factor.

As part of our performance measurement system, we used after-sales performance as one of the indicators of operational core process performance (see Exhibit B–3). We measured it using three criteria: sales growth of the after-sales service business, sales percentage of full-service contracts, and value-added per service employee. The survey revealed that successful performers in after-sales service can deliver 71 percent of their spare parts within 24 hours, while less successful performers manage only 36 percent. Successful performers generally carry out their service and maintenance assignments on schedule; only 11 percent are performed behind time. In comparison, less successful performers carry out 37 percent of all assignments behind schedule.

Successful performers in after-sales service can deliver 71 percent of their spare parts within 24 hours

But overall, IT use in after-sales service is still very limited—just under 50 percent even within the group of successful performers. The use of integrated standard systems support is also limited: 70 percent of successful performers in after-sales service use proprietary software and modified standard software, rising to 92 percent at less successful performers.

COMPETITIVE ADVANTAGES THROUGH IT USE

There are no substantial differences among the companies we surveyed in terms of the type of hardware used in after-sales service. At about 60 percent of all companies analyzed, service technicians use laptop computers. However, the companies differ widely as far as the use of hardware is concerned. At successful performers in after-sales service, service employees work on average about 2.4 hours per day with the laptop, which is twice as high as the comparable figure at less successful performers.

The reason for this difference lies in the functionality that the IT application systems provide (see Exhibit II–20). Successful performers in after-sales service use IT systems more intensively for personnel deployment, administration of service reports, and forecasting demand for spare parts—that is, for simplifying service tasks and improving the efficiency of the function.

But successful performers in after-sales service have also understood the value of IT for leveraging the effectiveness of service. They achieve impact by pursuing three main principles:

- **Give access to the right data about customers, products, and spare parts.**
  Data about spare parts and product specifications can easily be stored in databases, but even at successful performers in after-sales service, computerized availability of these data is relatively low (approximately

Exhibit II–20
IT-Based Functions in Field Customer Service

Professional service providers have more field service information electronically available

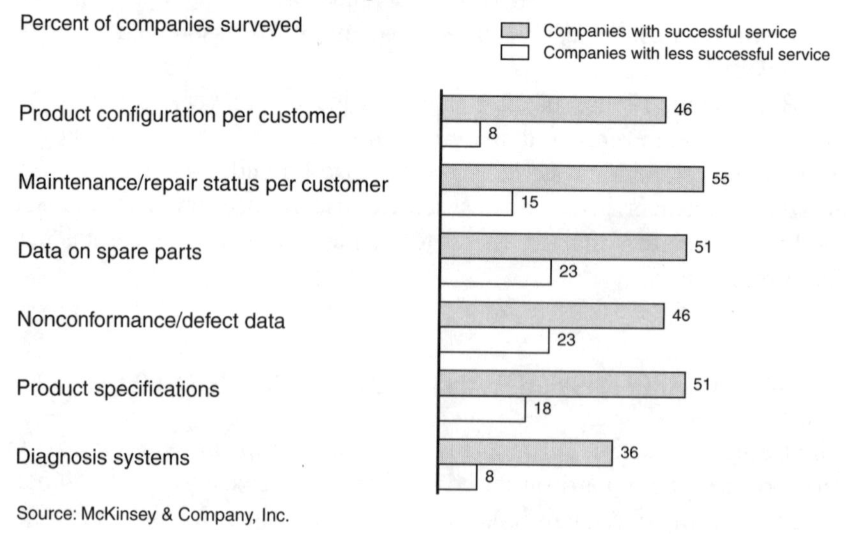

Source: McKinsey & Company, Inc.

50 percent). It is not the quantity of data that matters, but how usefully the data are processed for after-sales service. After all, the proof of how effective the information really is lies in its day-to-day utilization rate on the field service front. Stringent quality management is also essential; the data have to be current, consistent, and complete. If not, service technicians will simply continue working with copied lists, blueprints, and countless handwritten notes.

- **Learn from errors.**
  About 50 percent of successful performers in after-sales service work on the principle: "Errors are not a problem as long as you learn from them." They report errors systematically to the relevant divisions or functional areas—something only 20 percent of less successful performers do. But service technicians will not fill in error reports if they cannot be sure that these will be passed on to the R&D department or production and swiftly have an impact on existing or future products. It is the feedback that counts.
- **Simplify routine work via IT-based diagnosis.**
  One-third of successful performers in after-sales service support their service technicians with diagnostic systems that can identify standard errors and provide instructions for error correction. The technician intervenes only if the repair is particularly difficult. This substantially increases the service efficiency.

TREND TOWARD HIGH-TECH SERVICES

IT developments in the service sector for airplanes and railways are a good indication of future trends in this area. The main drivers will be the continuing price decline for microprocessors and declining costs of data transmission.

Service calls in the past were characterized by high traveling expenses and low productivity. The maintenance approach, especially of technology leaders, such as Trumpf and Heidelberger Druckmaschinen, has radically changed. Errors are increasingly being localized on-line. The service technician begins the error search by modem, and will conduct the repairs by remote control if possible. Even when there is no alternative but to visit the customer, on-line diagnosis offers detailed up-front information about the errors, and the service time is reduced—a benefit to both customer and supplier. This technology also offers growing opportunities for service providers to extend their service package (e.g., a global "24 × 7" service).

As plant and machinery are increasingly equipped with embedded

IT, the value-added of after-sales service is shifting toward software and programming technology. In many areas, the real value of service staff already lies in their knowledge of control software. Alongside actually implementing the systems, the main challenge for companies will be training to make sure their service teams have this new level of expertise.

SUMMARY

Our research shows that in the manufacturing sector, vast untapped potential for IT lies in marketing, sales, and service. Many best-practice examples indicate that quantum leaps in performance are possible if companies focus on the effectiveness of IT applications. The most successful companies use integrated base systems in their sales planning and order scheduling, and systematically support their maintenance and repair services with IT. Exhibit II–21 summarizes best practice in the three subprocesses we have examined and indicates future applications that are likely to have the most impact.

EARLY WARNING SYSTEM AND REMOTE DIAGNOSIS

COMPUTER MANUFACTURER

For some time now, a manufacturer of high-performance servers has been intensively using systems for the remote diagnosis and maintenance of their computers. In most cases it is not necessary to do the repair on-site at the customer. If the hardware needs to be replaced, early warning systems help to plan the work so that interruption of the customer's operations is minimized—for example, by delivering the required spare parts in advance and doing the repair work on weekends. The benefits of this IT application—such as lower downtime, reduced risk of data loss and breakdowns—have contributed tremendously to increasing the leading position of this manufacturer.

Exhibit II–21
Uses of IT in Marketing, Sales, and Service: Summary

	Status quo	Best practice	Outlook to the future
Marketing, sales planning	Using several IT systems, poor quality of data	Using integrated systems and data warehousing	Customer relationship management
Sales operations	Using several IT systems, poor quality of data	"Mobile office" with information vital for buying decisions, EDI, IT-based order checking	Product configurator, EDI via Internet, multimedia sales support
Customer service	Limited or no IT support	IT support for planning and execution, availability of databases, IT-based error detection, built-in checking via intelligent systems	Remote maintenance and repairs

Source: McKinsey & Company, Inc.

RULE 3

USE IT SELECTIVELY TO INTEGRATE ORDER PROCESSING ACROSS THE COMPANY

We have used *order processing* as a generic term to refer to two operational core processes: (1) materials and logistics management, with the subprocesses procurement and stockkeeping, internal logistics, and shipment and distribution; and (2) manufacturing. (Order intake using EDI was discussed in Rule 2.) The goals of order processing are short delivery times, high on-time delivery, low stocks, and high capacity utilization. These goals apply to both operational core processes under discussion: materials and logistics management, and manufacturing (see Exhibit II–22).

Obviously there is an inherent conflict among these four goals. The optimal solution has to be found depending on the type of business concerned.

To measure the performance of order processing, we used an indicator with three criteria: value-added per employee as a measure for labor productivity, on-time delivery performance, and progress in reducing throughput time in the period 1993 to 1996 (see Exhibit B–3).

Exhibit II–22
The Goals of Order Processing

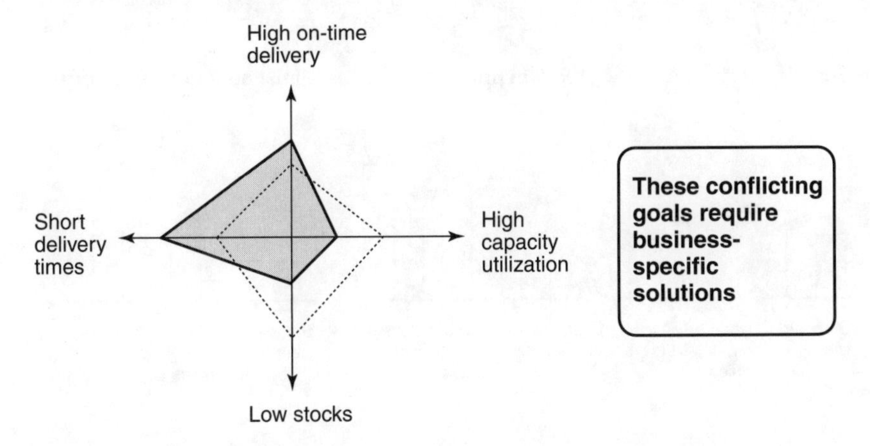

Source: McKinsey & Company, Inc.

IT APPLICATIONS FOR MATERIALS AND LOGISTICS MANAGEMENT

Our research in the manufacturing sector has shown that successful performance in order processing is primarily due to having much less working capital tied up in stocks. Without exception, the stocks of successful performers in order processing are on average 50 percent lower than those of less successful performers. Surprisingly, this is true for all types of stock—parts stock, work in process, and finished goods inventory (see Exhibit II–23).

Stocks of successful performers in order processing are 50 percent lower

Without IT support, companies cannot have complete command of procurement, purchase order processing, and inventory management. Consequently, the companies surveyed had high IT penetration in materials and logistics management, such as dedicated warehouse management systems or EDI-based materials ordering. The IT penetration of successful performers in materials and logistics management is 100 percent, as opposed to 83 percent at less successful performers.

Exhibit II–23
Typical Stock Rate versus Order Processing Performance

Companies with successful order processing manage with lower stocks

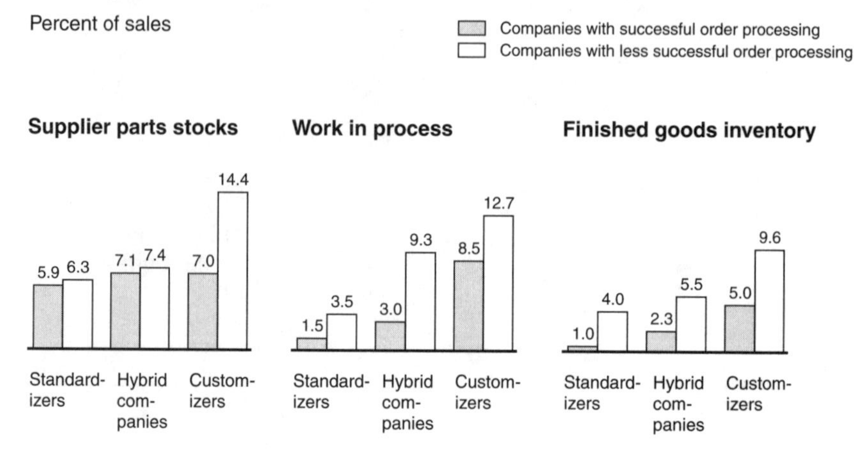

Source: McKinsey & Company, Inc.

The prime reason for using IT in materials and logistics management is to leverage efficiency. The introduction of bar coding was particularly important; it has helped to make performance in internal logistics, shipment, and distribution more transparent. This technique allows material flows to be controlled end to end without gaps.

Technically, it would be possible to give customers direct access to manufacturing and distribution information via the Internet so they could check on the status of their orders. But many companies are reluctant to do this; they prefer to maintain their freedom to bring forward or postpone orders without customer intervention.

Above all, successful performers in materials and logistics management achieve strategic advantage through a new form of supplier management. They have developed a privileged relationship with their suppliers, assigning first-tier suppliers the responsibility for choosing appropriate manufacturing technologies and conducting quality management. This means they invest more time in selecting suppliers and use IT to track the performance of suppliers continuously. For example, incoming shipments are no longer inspected routinely; instead, they are monitored via statistically defined samples.

The key benefits of successful supplier management are twofold:

- **Lower costs due to fewer supplier problems**
  More precise coordination between buyer and supplier reduces costs and avoids frustration for all parties involved. The costs of incorrect, incomplete, or untimely supplies are only 0.8 percent of purchase volume for successful performers in materials and logistics management. At less successful performers, the costs are 3.5 times as high.
- **More punctual supplies**
  Great potential could be unlocked by improving the punctuality of deliveries, even for successful performers. About 8 percent of their deliveries are delayed, in comparison to 12 percent at less successful performers. These poor figures have a negative knock-on effect: most companies take precautions by carrying higher stocks, which costs them more in working capital.

One important IT tool in supplier management is bar coding, combined with tracking and tracing systems. For expensive and time-sensitive goods in complex logistical settings, successful performers in materials and logistics management increasingly combine this technology with global positioning systems (GPS). These systems make it possible to track goods continuously in the transportation flow and improve the room for maneuver of manufacturing management. The combination of IT applications with GPS makes it possible to identify supply delays automatically

COMPETITIVE ADVANTAGES THROUGH IT-BASED MATERIALS AND LOGISTICS MANAGEMENT

AUTOMOTIVE SUPPLIER

Smart IT management of materials and logistics can play a vital role in gaining a competitive advantage. A supplier of bumpers for automobiles won the order to build a just-in-time production line near the customer. A key factor was the supplier's IT competence. The products are preassembled car body elements consisting of bumper and headlight that are produced in many varieties. This type of production requires enormous IT skills in materials and logistics management to coordinate processes at such short notice—the order lead time was only four hours—and to ensure low inventory and transportation costs.

and respond early. Now that the prices of GPS receivers are beginning to fall, the opportunities for application of this technology will increase, and it can be used in production processes for less expensive goods too.

Even more important than the individual aspects described so far is an understanding of the function of materials and logistics management as part of the process chain, consisting of the elements plan, source, make, and deliver. Only companies that activate every link in this process chain from their suppliers and suppliers' suppliers through to their own customers and customers' end users can successfully realize the benefits of supply chain management.

Building intercompany process chains poses high demands on IT systems and IT management. Success depends on the standardization of heterogeneous IT logistics systems—on whether it is possible actually to wire up the different companies. Partnerships with logistics specialists who manage the transportation flows and IT infrastructure between suppliers, manufacturing locations, and distributors can play an important role in this context.

For most companies—including those surveyed—this type of supply chain management is still something of a phantom, apart from a few companies in the "virtual enterprise" category. Their business strategy is to procure products tailored to the customer's needs by integrating different suppliers without any real net output from manufacturing facilities of their own. For these virtual companies, the logistics system is the connecting link between different plants and the crucial performance driver.

IT SUPPORT FOR MANUFACTURING

The use of IT in manufacturing, one of its traditional areas of application, has often been driven by excessive expectations. The idea of the computerized factory—the concept of computer integrated manufacturing (CIM)—remains largely a pipedream. But the issue still remains: What contribution can IT make to fulfilling the conflicting goals of order processing: short delivery time, high on-time delivery, low stocks, and high capacity utilization within the context of manufacturing (see Exhibit II–22)?

Even IT cannot square the circle, but which goal conflicts do companies with successful manufacturing solve better than their competitors? How do they use IT to optimize their manufacturing processes?

Today, production planning and scheduling (PPS) systems are widely used to organize purchase orders efficiently, monitor order processing, and plan materials requirements.

Exhibit II–24
Conditions for Using PPS Systems, by Type of Business

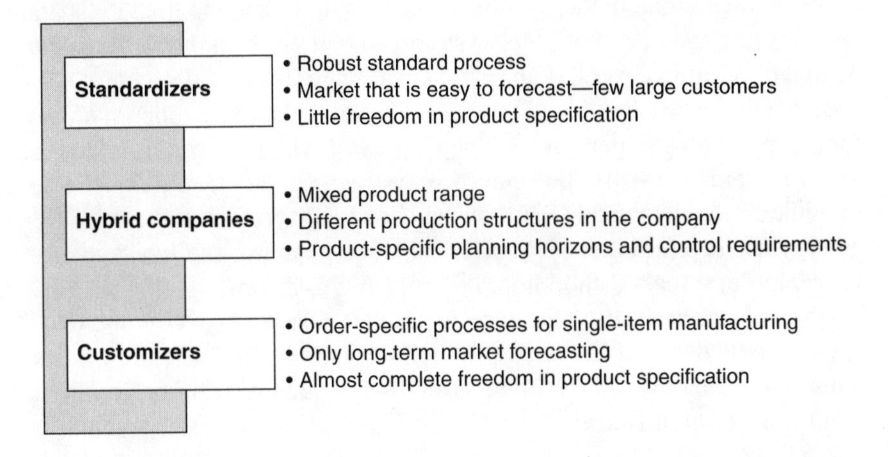

Conditions for using PPS systems vary according to type of business
Standardizers
• Robust standard process
• Market that is easy to forecast—few large customers
• Little freedom in product specification
Hybrid companies
• Mixed product range
• Different production structures in the company
• Product-specific planning horizons and control requirements
Customizers
• Order-specific processes for single-item manufacturing
• Only long-term market forecasting
• Almost complete freedom in product specification

Source: McKinsey & Company, Inc.

The choice of PPS system depends on the type of business. We find it helpful to think in terms of three categories (see Exhibit II–24):

- *Standardizers*—manufacturers of standard products (e.g., automotive suppliers, component manufacturers, large-scale manufacturers)
- *Customizers*—manufacturers of specialty products (e.g., single-item manufacturers, heavy capital goods industry, plant engineering and construction)
- *Hybrid companies*—manufacturers that operate both large-scale and single-item production

DEFICITS OF IT-BASED PRODUCTION PLANNING AND SCHEDULING

Two types of production planning and scheduling systems (PPS) are used for scheduling: materials requirements planning (MRP) and the currently most frequently used manufacturing resource planning (MRP II).

These types of software are often closely linked to IT applications in sales, procurement, and materials and logistics management. This IT system integration supports management in planning the manufacturing process. Inquiries from customers concerning delivery time can be answered by an IT system on the basis of calculated values, instead of rough estimates due to the coupling of production planning, materials and logistics management, and sales systems, so that potential bottlenecks can be identified before placing an order. Up-to-date integrated standard software combines MRP II features with sales and accounting functionalities, for example, on an open systems platform with client/server architecture. These systems are also called enterprise resource planning systems (ERP). In addition to dedicated MRP II systems or ERP systems, companies frequently use several other systems for detailed planning and fine-tuning—for example, so-called shop floor control (SFC) systems.

Most important, the survey revealed that despite the multitude of IT applications, users of IT-based production planning and scheduling systems expressed only limited satisfaction. We frequently heard critical comments (see Exhibit II–25).

Exhibit II–25
Criticisms of Production Planning and Scheduling Systems (PPS)

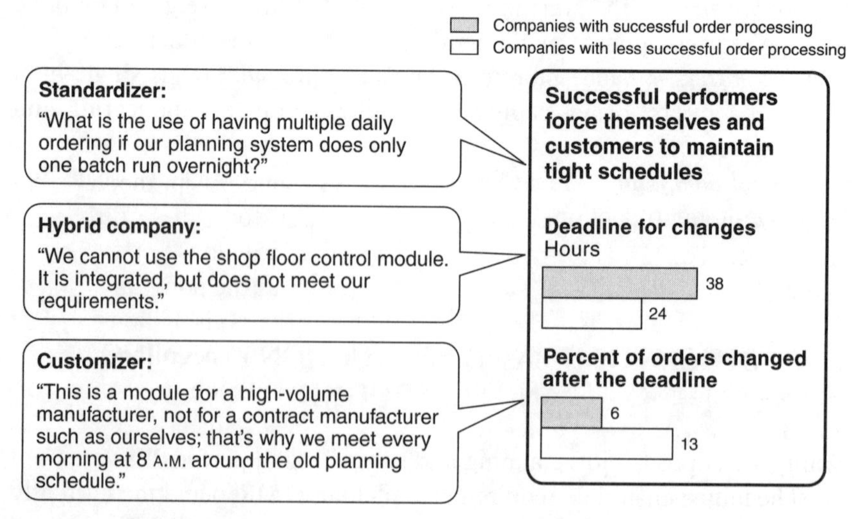

Source: McKinsey & Company, Inc.

We identified the following key deficits of PPS systems:

- Overcomplexity. PPS systems are too complex, so the systems functionality is not fully utilized.
- Inflexibility. System structures are difficult to adjust to the dynamic business processes.
- Conceptual weaknesses of the MRP II logic. Multistage planning first assesses materials requirements and then levels capacity requirements, with the consequence that the cost optimum cannot always be determined. A few software vendors try to escape this dilemma by installing further programs for simultaneous planning in addition to the basic PPS system, and they can make a difference. A U.S. steel manufacturer managed to increase output by 20 percent without expanding the plant's technical capacity.
- Throughput time syndrome. When planned schedules are not met, manufacturing planners raise the planned values, which merely advances the start time of all orders and does not get to the crux of the problem.
- Lot size difficulties. Calculating lot sizes from a cost viewpoint tends to result in very large lot sizes. As a result, the schedules of other orders are pushed back. In practice, this is often corrected manually by dividing big lot sizes into several small ones to lower stocks and throughput time.

So how do successful performers in order processing succeed in using PPS systems successfully?

BEST PRACTICE FOR PRODUCTION PLANNING AND SCHEDULING (PPS)

Successful performers in order processing know and accept the technical limits of their PPS systems—and manage to transform these limits into advantages. The key is the culture: the disciplined way in which staff work within the given constraints. The order freeze point (the point up to which changes in orders are possible before production starts) is 38 hours at successful performers in comparison to 24 hours at less successful performers (see Exhibit II–25). Successful performers strictly observe this self-imposed rule. After the order freeze point, only 6 percent of orders are changed, as opposed to 13 percent at less successful performers.

The touted customer orientation of less successful performers in order processing causes higher manufacturing costs, which customers rarely actually pay for. Moreover, control of the manufacturing process becomes so difficult that agreed deadlines for delivery often overrun.

Amazingly, successful performers in order processing rely primarily on the competence of their staff to solve manufacturing problems, whereas less successful performers rely on greater use of IT to control increasing complexity. Frequently hybrid companies and customizers try to steer the production process via expensive, large-scale IT support. But these efforts often fail precisely because greater complexity is all the more prone to human error and chance occurrences.

A good example of manpower versus IT power is the organization of fine-tuned planning in production—so-called shop floor control (see Exhibit II–26). Seventy-eight percent of successful performers in order processing manage shop floor control without IT support, compared to 34 percent at less successful performers. To optimize the operational level, these companies rely on established methods of structuring operations—for example, Kanban systems, cumulative timing, or work in groups.

78 percent of successful performers in order processing manage shop floor control without IT support, compared to 34 percent at less successful performers

Exhibit II–26
Methods of Shop Floor Control

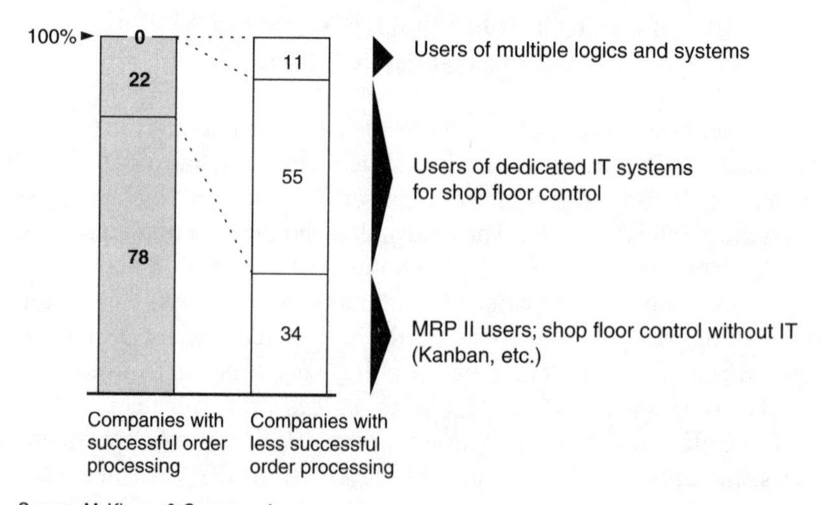

Source: McKinsey & Company, Inc.

Exhibit II–27
Strategies for Implementing PPS Systems

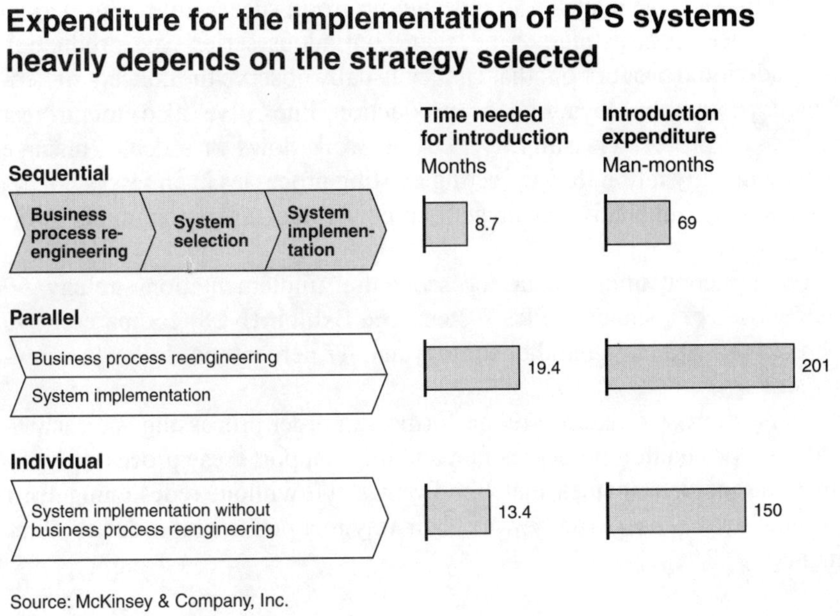

Expenditure for the implementation of PPS systems heavily depends on the strategy selected

	Time needed for introduction (Months)	Introduction expenditure (Man-months)
Sequential: Business process re-engineering → System selection → System implementation	8.7	69
Parallel: Business process reengineering / System implementation	19.4	201
Individual: System implementation without business process reengineering	13.4	150

Source: McKinsey & Company, Inc.

Differences in PPS implementation strategies also spotlight high performers' preference for organizational solutions and their skepticism of unreflected use of complex IT in this area (see Exhibit II–27).

There are three basic approaches to implementing a PPS system:

- **Sequential implementation**
  Implementation begins with a redesign of the production process. A suitable PPS system is selected based on the requirements of the new process. The PPS system is then implemented as the third step.
- **Parallel implementation**
  Process optimization takes place simultaneously with implementation of the PPS system.
- **Individual or separate implementation**
  The PPS system is implemented without any redesign of the production process.

Successful performers in order processing generally prefer the sequential implementation approach. Good preparation in the form of process optimization pays off: companies that pursue this strategy usually have much lower implementation costs and find the implementation itself is faster.

In making the decision to redesign up front, successful performers in order processing distinguish between continuous series-type production and individual production; that is, they usually observe the concept of dedicated production locations or production lines. We also found that successful performers simplify existing work flows in order to manage them manually rather than reflecting existing processes in an IT system. To do this, they frequently rely on staff empowerment and autonomous working groups.

When companies pursue the sequential implementation strategy, 60 percent select a standard PPS system (see Exhibit II–28) compared to 40 percent that pursue a parallel strategy and 47 percent that follow a singular strategy.

The strategy of successful performers in order processing is clearly to redesign production processes first and then support these processes by IT applications. Companies that blindly trust IT without redesigning their production processes rarely work their way to a position of industry dominance.

Exhibit II–28
Benefits of Various Strategies for Implementing PPS Systems

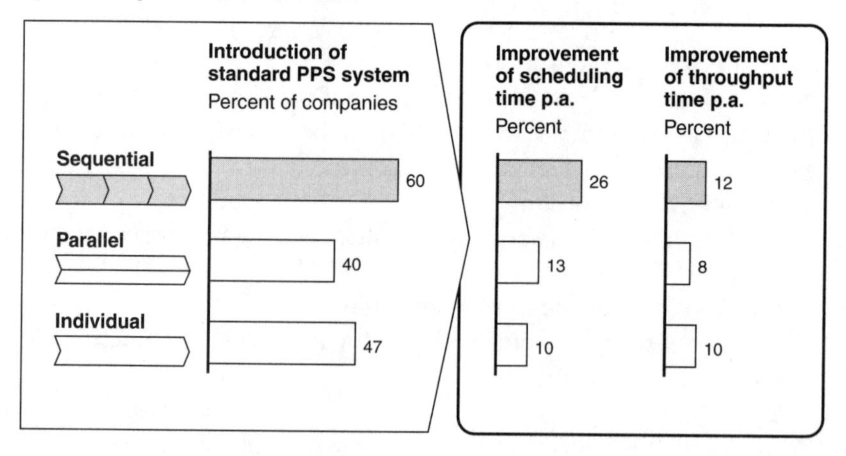

Source: McKinsey & Company, Inc.

These factors for success are likely to remain valid for some time to come, as no revolutionary new developments in the base functions of PPS systems appear to be on the horizon.

Currently, the major software vendors are primarily focusing on details and supplementary functions in the field of analysis tools and planning support. For example, "available to promise" applications help to plan orders while taking into account available capacities and materials simultaneously, covering work in process and stocks of finished goods as well as feedstock.

New analysis programs allow use of the PPS data to evaluate alternative production scenarios, relevant when a company that is working to capacity must decide whether to accept an additional order. This raises questions such as: "Which confirmed orders would be delayed?" and "How long would the delay be?"

These new developments are characterized by excellent graphic presentation of the complex facts. This enables the production planner to identify conflicts and immediately take corrective measures. Intelligent software systems reduce the volume of calculations during simulation through special calculation techniques like upstream and downstream propagation (i.e., analyzing the impact of scheduling changes at the upstream and downstream production stages).

Recent planning software no longer sticks to the common logic of PPS systems. According to conventional wisdom, these try to determine the cost optimum stepwise across a forecast of primary demand, materials requirement planning, and production scheduling. Beyond costs, the new systems take into account parameters such as on-time delivery and inventory management.

New PPS systems that support supply chain management have broken away from common MRP II optimization methods, which tried to solve capacity bottlenecks related to one manufacturing unit. These so-called local methods do not sufficiently take into account the total throughput of a manufacturing facility because other manufacturing units are not considered (and so their capacity remains unused). New intelligent algorithms in combination with methods for constraint-based planning provide a remedy by global optimization of production. This new approach determines the maximum throughput, considering materials and production resources simultaneously.

Beyond intensive IT support for production planning and scheduling, there are numerous additional areas of application for IT in manufacturing—for example, plant layout planning, assembly simulation, assembly support, and IT support for automation of manufacturing (e.g., welding

IT SUPPORT FOR ASSEMBLY WORK

CABLE LAYING AT BOEING

The U.S. airplane manufacturer Boeing is working on one of the first virtual reality applications in manufacturing. In a pilot test, employees wear a mini-computer on their belts and a headset with a camera and a transparent computer screen in front of one eye. The camera automatically logs the measurement points from the assembly walls in the plane. The computer then uses these data to determine the exact position of the employee in the fuselage at any time. Then the system scans the instructions for cable laying onto the transparent screen. Remarkably, this new method reduces assembly time by almost 50 percent—and that does not take into account the better quality of cable laying and the lower error rate.

robots, NC/DNC machines). In the more distant future, areas like automation and robots, IT-based management of work flow documents, and conventional planning and scheduling systems will link up to or merge with more complex IT environments, opening up the potential for radical changes to the manufacturing process. The example of cable laying at Boeing gives an outlook to the future.

SUMMARY

Nearly all the companies we surveyed have installed comprehensive operational systems for purchase-based order processing, stockkeeping, and manufacturing resource planning, but the less successful performers tend to overrate IT-supported control at the shop floor level. Successful manufacturers, in contrast, often find it is more effective to use simple organizational processes like Kanban and limitation of their customers' order scheduling flexibility. Maintaining control of internal order processing will become even more important if the supply chain becomes more integrated in the future, with IT providing more direct links to customers and suppliers. Exhibit II–29 summarizes status quo, best practice, and outlook for the key areas concerned.

Exhibit II–29
Uses of IT in Order Processing: Summary

	Status quo	Best practice	Outlook to the future
Materials and logistics management	IT-based purchase order processing, stockkeeping	IT-based inventory management, quality management	Supply chain management
Manufacturing	MRP II	IT-based production planning/shop floor control without IT	Improved optimization algorithms for capacity planning, simulation models

Source: McKinsey & Company, Inc.

RULE 4

SHIFT THE FOCUS OF IT IN ADMINISTRATION TO BUSINESS PLANNING AND MANAGEMENT DEVELOPMENT

Is the use of IT in administration really a relevant subject for executives? Hasn't the field already been saturated with standard software that can automate and align administrative processes across all sectors of industry? In many areas of administration, IT applications have reached diminishing returns, but a fresh perspective on the scope of IT in administration is needed.

The structural changes affecting every facet of industry have long since shifted the coordinates of administrative activity:

- Globalization of all output and input markets is intensifying competition but also offers attractive opportunities for expansion.
- The dynamics of demand preferences and technological development are increasing, combined with the drive toward more systematic orientation and greater product/process flexibility.
- The ongoing transformation in organization structures and values has empowered staff, pushing accountability down the ranks to operational units, and performance requirements have risen.

In today's increasingly unstable environment, administration should move beyond conventional processing of data from accounting and personnel to incorporate a new understanding of corporate planning, systematic management development, and effective knowledge management. This new approach to administrative functions implies changed demands on the supporting IT systems. IT in the past primarily improved the efficiency of administrative processes through automated processing of large data volumes. It now has to become the driver of effectiveness by providing more precise and intelligently processed data for administration's broader mission.

IT in administration has to provide more precise and intelligently processed data for administration's broader mission

In our examination of the use of IT in administration, we did not use operational metrics, as we did for the operational core processes, because the effectiveness of administration cannot be measured in a reliable and valid manner using indicators based on anything less than the company as a whole. As far as the indicator of administrative performance is concerned, we refer to the indicator of corporate success (see Exhibit B–1).

STILL THE PRIMARY FOCUS AT MOST COMPANIES: BASIC PROCESSING OF DATA FROM ACCOUNTING AND PERSONNEL

According to conventional wisdom, administration primarily has the function of replicating all of a company's transactions as data and processing these data efficiently into information. The responsibility for these administrative processes lies with accounting and human resources management. The operations in these departments are heavily repetitive and to a great extent uniform across all sectors of industry. It is therefore easy to automate these operations. Consequently, administration was the first business function in which IT systems succeeded in winning widespread acceptance. In the view of IT users, the benefits were obvious: fast and cost-effective processing of large data volumes reaped impressive efficiency gains.

As a result of the early efficiency-driven spread of IT in administration and the accompanying system standardization, today all companies surveyed use comparable basic software. Around ninety-five percent use standard software in accounting, of which two-thirds use integrated software packages such as SAP R/3 or similar products (see Exhibit II–3). About eighty-five percent of the companies surveyed use standard software programs in human resources. However, these are generally not integrated packages, but area-specific tools such as Peoplesoft applications. The reason is that from a viewpoint of data organization, integration with other functional areas offers comparatively low benefits.

The penetration of individual operations in administration is also higher than in other functional areas. In the companies we analyzed, more than 85 percent of employees in administration, on average, have computerized workplaces, with access to the various IT systems, as well as office applications such as word processing and spreadsheets.

Does this mean the remaining potential for improvement through IT in administration is low, because most companies have achieved acceptable minimum standards for systems and work flows? The answer is definitely no, as the following results indicate.

CURRENT DIFFERENTIATION OPPORTUNITY: SUCCESSFUL COMPANIES ACHIEVE TWICE THE PLANNING QUALITY AT HALF THE COST

We found that financially successful companies spend an average of only 1.4 percent of revenues on controlling, accounting, and human resources, whereas less successful companies spend 2.7 percent. More important than this are the figures on planning quality. Less successful companies estimate their trend in return on sales with an average precision of ± 4.0 percentage points, whereas successful companies achieve an accuracy of ± 1.7 percentage points (see Exhibit II–30).

This weakness in planning accuracy can have damaging repercussions on a business. Against the backdrop of increasingly efficient international financial markets and intense debate on shareholder value, a negative deviation can quickly be interpreted as a risk. Emerging liquidity problems, loss of confidence among institutional investors interested in optimal performance of their portfolio, and declining credibility of management can draw the company into a downward spiral.

Even the rare case of a positive deviation can have a counterproductive effect. Overly conservative planning implies that crucial investments are not being made (or only half-heartedly)—that the company is not seizing market opportunities.

So how do financially successful companies manage to achieve a planning quality twice as high at half the costs, although they mostly use standard software available to all competitors? Their decisive lead in efficiency

Exhibit II–30
Administrative Expenses and Deviation from Plan

Successful companies have efficient administration and more effective controlling

Percent of sales revenues

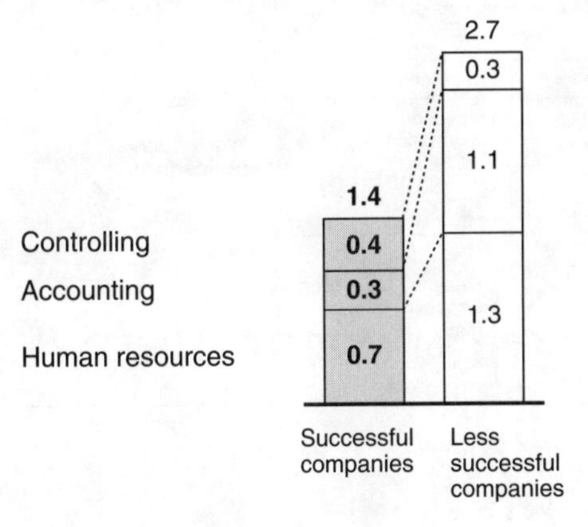

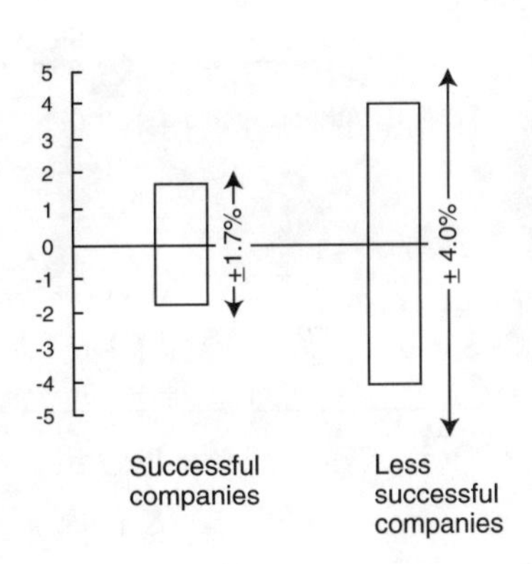

Source: McKinsey & Company, Inc.

is due to the greater integration of their basic IT systems, especially the IT-based link of their administrative functions with order processing. Financially successful companies integrate their administrative systems with IT applications used in other functional areas almost twice as frequently as less successful companies, and achieve much higher availability and consistency of data. At the same time, they reduce nonproductive activities to a minimum. Financially less successful companies often spend more than half their total working hours on these activities.

Their lead in effectiveness largely results from a broader understanding of what administration involves and the benefits offered by IT. At top performers, administration covers much more than simply handling data from accounting and human resources. Controlling is not confined to planning and monitoring operational and financial ratios; it also directly supports the company's strategic planning. Top performers have a clear focus on creating value-added to strengthen the operational capability of other functional areas of the company. As well as processing basic data, they offer information architectures with decision-supporting functions (see Exhibit II–31). They use work flow management applications, add-on

Exhibit II–31
IT Applications for Administration

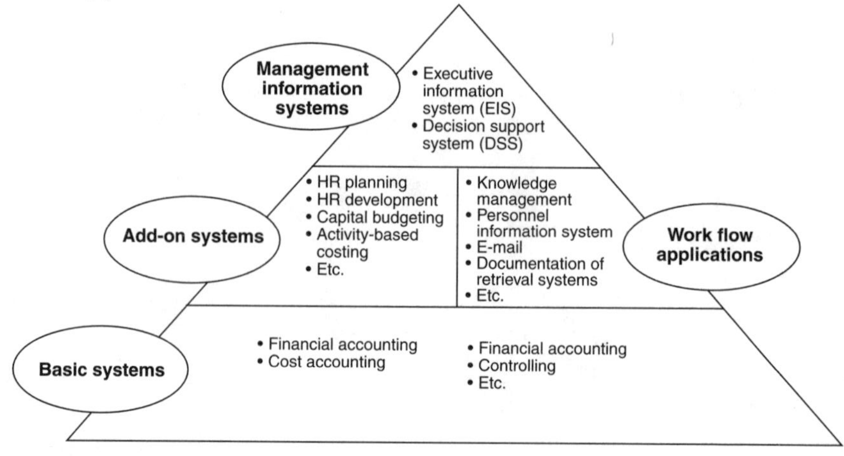

Source: McKinsey & Company, Inc.

PERSONNEL INFORMATION SYSTEM USING LOTUS NOTES

U.S. ELECTRONICS MANUFACTURER

A U.S. electronics manufacturer uses a personnel information system based on Lotus Notes to provide its staff with a constant stream of fresh information. The system rates each piece of information on a scale by its relevance to the user and can mail the data to defined target groups (see Exhibit II–32).

All types of information—external and internal, analog and digital—can be included in the system. Analog information is digitalized by scanning. So-called gatekeepers are responsible for looking through defined information media—for example, magazines, on-line information services, and customer newsletters. They feed relevant contributions and data into the information system and rate each piece of information according to its relevance.

All employees with access to Lotus Notes can specify a reader profile, defining whether and in how much detail they wish to receive information on selected subjects. The information is distributed by the system daily to each recipient's mailbox. Users can adjust their profile or mailing list at any time.

Exhibit II–32
Personnel Information System of a U.S. Electronics Manufacturer

Quick and user-friendly provision of information

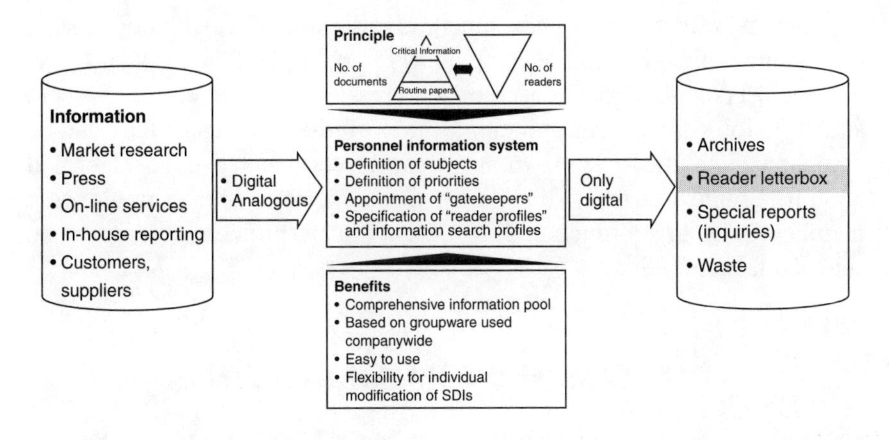

Source: McKinsey & Company, Inc.

systems for accounting and human resources, and management information systems much more frequently than less successful companies do.

WORK FLOW MANAGEMENT APPLICATIONS

Simple work flow management systems like e-mail are used much more frequently at financially successful companies. At successful performers, two-thirds of administrative staff send and receive e-mails; this figure is under 50 percent at less successful companies. Many successful companies have also implemented intelligent IT applications that build on the basic functionalities of work flow management systems.

Such work flow management systems accelerate and simplify information processes. Usually they are implemented not only in administration but in all other functional areas of the company too.

ADD-ON SYSTEMS FOR ACCOUNTING AND HUMAN RESOURCES

Supplementary IT applications extend the functionality of conventional IT systems in accounting and human resources (HR) (see Exhibit II–33). Financially successful companies use these applications primarily to improve the effectiveness of administrative functions. For functions such as personnel planning and development, for example, they frequently use dedicated IT applications (based on standard PC applications) to process and store job descriptions and employee profiles. This means they can be compared, and the type of training employees need can be identified quickly. Employees most suitable for a specific task—for example, as a member of a cross-functional or international project team—can be selected fast and reliably.

MANAGEMENT INFORMATION SYSTEMS

Successful companies use IT systems such as the EIS module from SAP or HYPERION to provide relevant information for top management, control-

Exhibit II–33
IT Support for the Human Resources Function

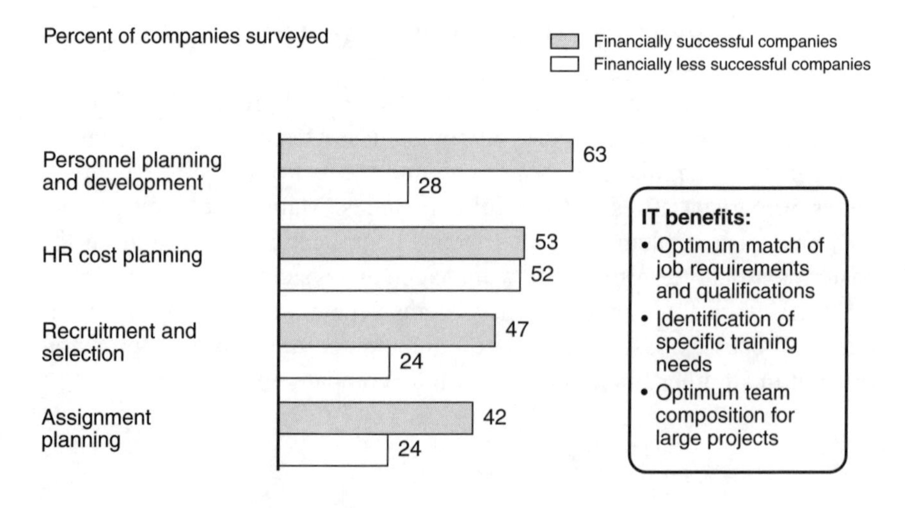

ling, and other functional areas quickly and tailored to the needs of the recipient. They rigorously focus on exception reporting, and offer support for causal analysis.

Key indicators are available at the touch of a key twice as frequently at more successful companies

The best companies already use flexible management information systems (MIS) with a simple user interface for top management. These systems support multivariate analysis. Important indicators such as profitability by division or by customer are available at the touch of a key twice as frequently as at less successful companies (see Exhibit II–34).

Today standard management information systems can be adjusted flexibly to the needs of different users. However, for successful companies, comprehensive information design is essential for systems implementation. The individual information requirements of decision makers in different functional areas and at different hierarchical levels must be identified and specified before the system is rolled out.

During the design process, it is important to standardize definitions of the relevant management indicators (such as return on investment, stock turnover, value-added per employee, and sales per sales representative). It

Exhibit II–34
Superior MIS: Characteristics of Successful Companies

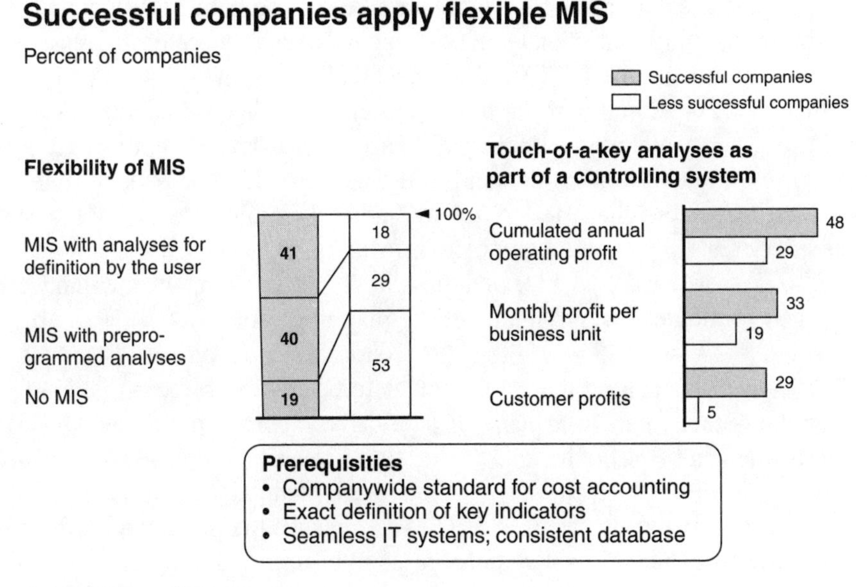

Source: McKinsey & Company, Inc.

is also necessary to specify target groups, type and frequency of data collection, as well as the editing and visual presentation that will be required. At successful companies, this comprehensive information concept is an integral part of the MIS.

Successful companies see the design process as an iterative process, where planning and control systems are adjusted continuously to changing conditions. They review their ratio systems regularly to check the relevance of indicators, devising new ones where necessary and eliminating those that are no longer relevant. For example, training days are no longer checked if the training program itself has been completed successfully.

This knowledge about which indicators are really important is a prerequisite for effective IT support of top management and controlling.

FUTURE PERSPECTIVES: IT SUPPORT FOR BIFOCAL BUSINESS PLANNING

Development of a company's communication infrastructure (e.g., work flow management systems) can leverage performance in administrative processes dramatically. Numerous digital document management and retrieval systems, video and computer conferencing systems, and Internet applications are already available. These systems can accelerate information exchange both internally and between companies and their business partners.

But the greatest lever for success is likely to be better IT support of business planning and management. In spite of the generally impressive best-practice examples of IT stars, almost all IT users have obvious deficits in this area. For example, so far few IT systems have been available to support long-term planning effectively. Apart from a few exceptions (e.g., determining medium-term demand in the automobile industry), complex forecasting systems have failed due to market dynamics, the multitude of boundary conditions, and incomplete information.

To improve IT support for planning, in the future companies will have to adapt to the new requirements of the planning process. The debate on shareholder value already shows a clear trend toward bifocal planning: data-driven short-term planning (one to three years) and vision-oriented long-term planning (five to ten years). In short-term planning, the role of IT is to improve the precision (defined quantitively) of planning. The purpose of long-term planning is to transform pure financial planning into strategic planning that is oriented to

Short-term planning has a more quantitative focus; long-term planning is oriented to qualitative goals

Exhibit II–35
IT Requirements of Bifocal Planning

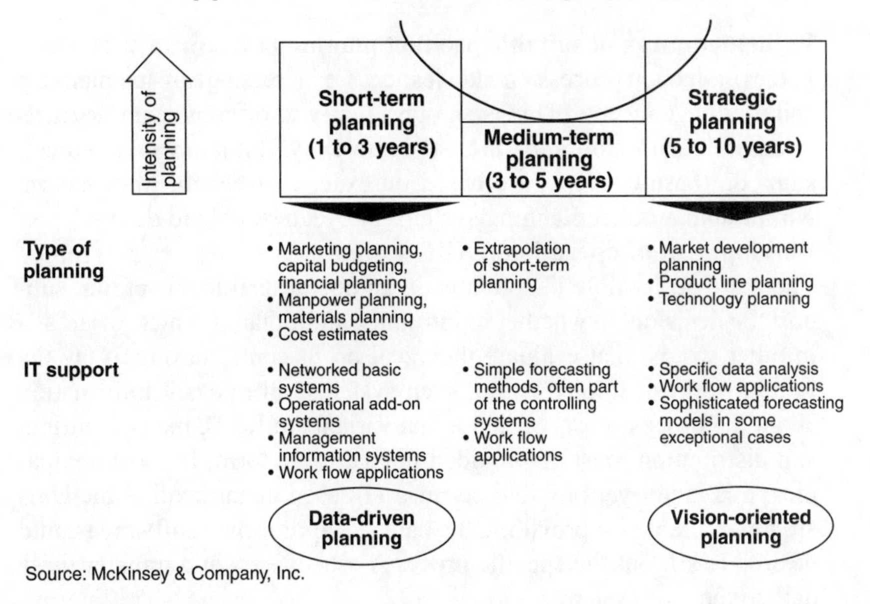

Source: McKinsey & Company, Inc.

qualitative goals. Medium-term planning (three to five years) provides the link between short-term and long-term planning (see Exhibit II–35).

Bifocal planning attempts to do justice to changing market conditions by concentrating on both the short and the long term. Increasingly complex and unstable market conditions are forcing companies to evaluate future opportunities and risks as concretely as possible at an ever earlier point in time. Under the growing pressure of international capital markets, they have to develop clear long-term scenarios on how they can continuously enhance shareholder value, and adapt their business policy accordingly.

IT SUPPORT FOR SHORT-TERM PLANNING

Short-term planning can capture significant productivity potential by improving the utilization and evaluation of all the data available in base IT systems. The following examples illustrate how this can be done.

ESTABLISHING ACTIVITY-BASED COSTING

Accounting must become more flexible and transparent so that the cost drivers of critical business processes can be identified and allocated to their true cause. This is an essential lever of complexity management. The key way to do this is to install systematic activity-based costing (ABC). Following are two examples:

- **Implementation of suitable product planning systems**
  In the production process, market leaders are increasingly using platform and module concepts. This is an opportunity to offer as many variants as possible with minimum internal complexity. But it is vital that decisions on these concepts are based on exact cost-benefit information. No suitable product planning systems are yet available to do this.
- **Development of operational ABC systems**
  There are no feasible IT systems on an operational level yet that support the decision on whether to introduce additional product variants—in other words, that evaluate their real profit contribution. To lay the groundwork for implementing such systems, all relevant information about the process costs of individual variants in R&D, manufacturing, and distribution must be provided in electronic form. In most companies, this is not yet possible because IT-based quantification methods are not available for providing the data. An appropriate software should be able to sift out the specific process costs exactly and transfer these data to the cost system.

TRANSITIONING TO DECISION SUPPORT SYSTEMS

Priority should be given to developing existing MIS systems into comprehensive decision support systems (DSS). Considerable qualitative improvements can be achieved by using new integrated database concepts and automating data analysis more rigorously:

Existing MIS systems need to develop into comprehensive DSS systems

- **Data Warehousing**
  Unlike many MIS systems, real-time data warehousing allows the user to access and process data from engineering and quality management systems, in addition to financial data. This resolves many complaints about typical deficiencies of MIS (e.g., insufficient currency of data and excessive focus on financials).
- **Automated data analysis**
  This allows a multitude of user-driven analyses such as ABC or portfolio analysis, but also data-driven analyses, such as the search for data discrepancies and outliers (including causal analyses). Frequently the

software has an "assistant" available, so that one-off problem-driven analyses can be conducted at minimum cost.

IT SUPPORT FOR LONG-TERM PLANNING

A characteristic of vision-oriented planning is that it combines the best elements of strategic planning (e.g., market segment planning, dynamic industry structure analysis, scenario techniques). These tools can be used to develop annual targets, action plans, and budgets by "rolling back the future"—starting at a given point in the future and working backward. The required target capabilities of the company must be identified and their systematic development planned. The goal of the planning process is to overcome barriers in thinking, establishing a mind-set that is proactive and able to anticipatc discontinuities.

The IT contribution inevitably focuses largely on indirect decision-making support (better, more target-group-oriented editing of information and faster, more precise supply of data). Of course, strategic long-term planning also includes the midterm base data provided via IT. Targeted data evaluation and the use of IT-supported forecasting processes can also be useful for special tasks, such as market analyses. But IT support clearly yields the greatest benefits in the field of organization, where new IT solutions, such as work flow systems, offer decision makers much more effective tools for coordinating and integrating planning tasks (similar to their importance in product development).

SUMMARY

The administrative function is transitioning from that of data manager into an internal service provider that improves the operative capability of other functional areas. This will fundamentally change the requirements for IT support in administration.

Up to now, IT has been primarily an efficiency driver. In the future it has to become an effectiveness driver. To supplement existing standard software for this purpose, it will be necessary to develop further work flow management applications, develop MIS systems into comprehensive DSS systems, and implement ABC systems. The objective should be to support short-term operational planning more effectively. For vision-oriented, long-term planning, effectiveness can be greatly enhanced by using IT-based knowledge management and special forecasting systems to back up strategy decision processes.

Part III

SUCCESSFUL IT MANAGEMENT

In Part II, we saw that if the right levers are applied, IT can greatly support the effectiveness of manufacturing companies' core business processes. But to have this impact, professional IT management is essential. This begins at the very top: successful IT management is definitely a top management affair. IT continues through to the IT organization, which needs careful structuring to be able to align the multiple requirements with best-practice criteria. And it is reflected in superior implementation management, always ensuring that business processes are redesigned before introducing the IT to support them.

We define the scope of IT management as all tasks involved in the planning, implementation, and operation of IT application systems (see Exhibit III–1). In addition to all planning and control activities, from procurement of hardware and software through network components and services to coordination of internal and external service providers, it also covers internal IT-related consulting. Although it encompasses IT systems development and operation, who actually carries this out varies from company to company. Sometimes the services are performed by the company's own IT management, sometimes by company-owned IT service units, and sometimes by specialized external service providers.

Why do companies have difficulty establishing successful IT management despite their concerted efforts? Our research on IT organization and management of IT projects showed the following key problems and weak points of IT management in the manufacturing sector:

- Often companies take too little time or lack the required skills to specify the requirements of IT users with sufficient clarity.

Exhibit III–1
Functions of Professional IT Management

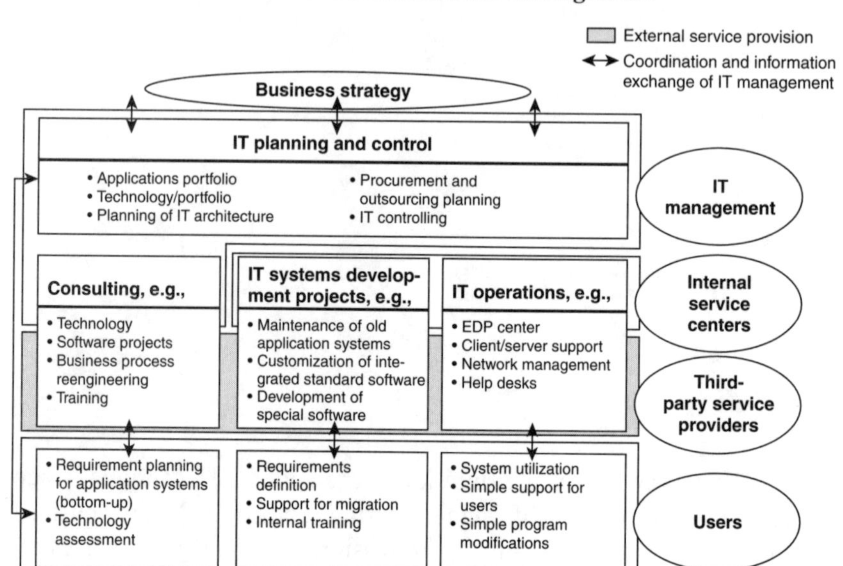

Source: McKinsey & Company, Inc.

- Companies create unnecessarily high complexity if they neglect to design a comprehensive concept for IT systems architecture, and then leave it to user departments to tap their IT support decentrally. It is no wonder that chaos ensues.
- Particularly companies with small IT departments can barely keep pace with the extremely short technological innovation cycles of vendors. Faced with the multitude of IT systems on offer, they may well lose track and give up.
- The replacement of old technology platforms, such as mainframes or proprietary software, often gets out of control. The scope and complexity of these projects usually require more than companies can handle with limited IT resources.
- Frequently companies respond far too quickly with outsourcing to constantly increasing IT demands, whether security management, systems availability, or user friendliness. This is especially the case when their own IT organizations are no longer able to meet their high service expectations.
- Often the manager of the traditional EDP department is put in charge of IT management. If neither the management tasks nor responsibili-

ties are clearly stated, the role is generally overly demanding, and the manager is unsuccessful.

IT stars demonstrate that there is another way. Their superior IT performance is based on top management attention, clear IT organization structures, and successful management of IT systems implementation.

RULE 5

MAKE IT A TOP MANAGEMENT AFFAIR

You would have thought that IT had long since made the top management agenda—if only because of its exorbitant costs and the constant need to upgrade systems. But involvement of top management in IT is by no means automatic, as the results of our survey show. We asked to what extent board members and the first layer of management concern themselves with IT issues. We found that at IT stars, top management spends an average of about 45 hours per month on IT, compared with 20 hours per month for laggards (see Exhibit III–2).

A limited number of board members should deal intensively with IT issues

The differences are even greater if you examine what lies behind these figures. On average IT laggards assign five senior executives IT management tasks, so that each devotes only around 4 hours a month to IT. At IT stars, on the other hand, on average only three senior executives deal with IT issues, each spending about 15 hours a month on IT issues—about four times as much as their peers at laggard companies. The advantage is obvious: 15 hours is enough time to become familiar with the real issues at hand, clarify questions in context, and—most important of all—align the use of IT to corporate strategy. In 4 hours a month, laggards can only perform routine activities such as budget control or project audits. The key to success is simple: only a limited number of top managers should deal with IT issues, but with sufficient intensity.

Not surprisingly, top managers at low-performing companies who spend an average of just 4 hours per month on IT tend to have vague and

Exhibit III–2
Top Management Involvement in IT Issues

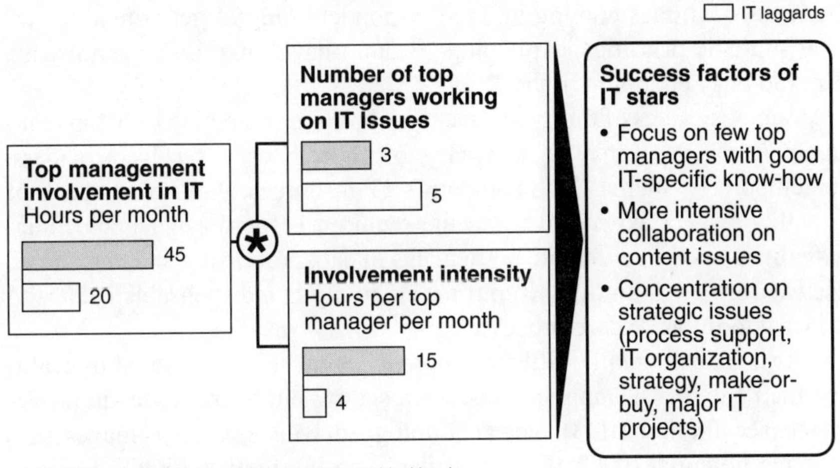

* Top management: board plus next hierarchical level.
Source: McKinsey & Company, Inc.

unrealistic expectations. These range from the over-ambitious ("cutting IT costs in half through outsourcing," wrote one) to the gloomy ("new media and IT tools are unsuitable for heavy engineering," said another). At such companies, business processes are seldom linked to IT goals, and even major IT projects receive scant top management participation.

ALIGNING IT STRATEGY TO BUSINESS STRATEGY

Involving top management is the best way to ensure that IT strategy is aligned to the business strategy. Here, too, IT stars set standards. At almost one in every two IT stars, IT objectives are derived direct from the business plan; at one in three, the objectives are set jointly with top management. This is the only way to guarantee that IT requirement profiles of the different user departments and the EDP department are aligned to the strategic orientation of the company.

Involving top management is the best way to ensure that IT strategy is aligned to business strategy

Some companies might argue that best-practice examples are all very well in theory, but reality is a different story. Instead of deriving IT projects directly from their business plan, companies tend to focus on matters that appear more pressing. Operational considerations often demand that priority be given to projects with lower strategic importance to make sure business continues coming in. One respondent did not get around to setting strategic priorities, explaining, "I don't have time to sharpen my ax. I'm too busy chopping timber."

Vicious cycles of this type have to be broken, something only top management can do. It may be necessary to call in external service providers, or temporarily increase the company's IT resources. A detailed check of monthly project expenditure and the required IT performance and functionality can often eliminate bottlenecks in data processing and help create leeway for new initiatives. Room for creativity is indispensable if IT is to be used to improve core processes.

Two fundamentally different approaches must be integrated to evaluate the benefit of IT applications when setting out to make this improvement (see Exhibit III–3). The **technology-driven approach** makes sure that the potential of IT already available on the market—both hardware and software applications—is used. This is especially important because

Exhibit III–3
Approaches to Measuring Performance of IT Applications

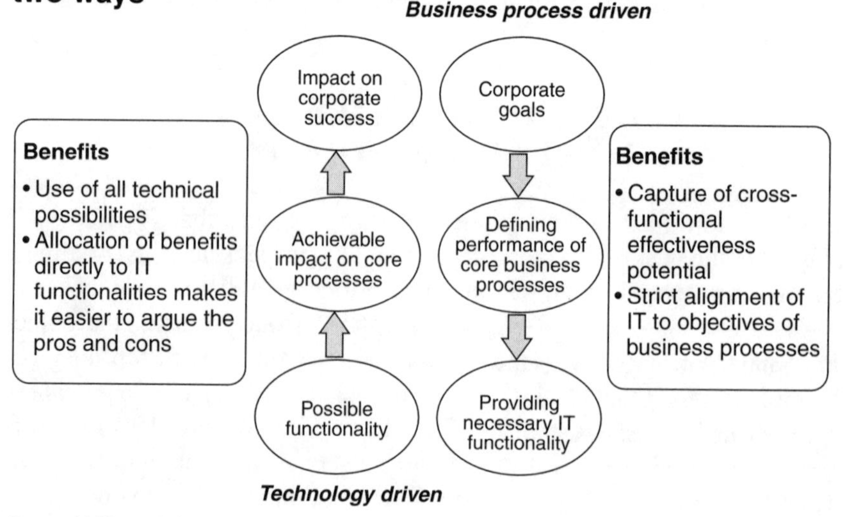

Source: McKinsey & Company, Inc.

IT is characterized by extreme dynamics of supply. Only continuous screening of the market ensures nothing is overlooked that could improve a company's bottom line. The **business-process-driven approach** is to select and implement the appropriate IT support only after identifying thc performance needs of core business processes. This makes sure only the essential functionalities are covered, leaving out all the unnecessary extras from the start. The integration of both approaches requires the cooperation of IT users and the IT department. Together they have to ensure alignment of IT strategy to business strategy in the course of joint IT planning. Top management has a crucial role to play in coordinating and guiding this process.

SETTING CLEAR OBJECTIVES

Today concepts like professional project management and the development of internal customer-supplier relationships are almost second nature, but it is very difficult to bring these approaches to life in an existing IT organization. Attempts to implement them will fail without precisely defined objectives and metrics to determine how far they have been fulfilled.

IT stars demonstrate how important it is to formulate tangible objectives with an exact time frame. They set specific objectives—for example, "We're going to increase the success rate of bids from 32 to 45 percent within two years," or "We will expand intranet use in the sales centers from 30 to 80 percent within six months." IT laggards generally come nowhere near this level of detail. They make do with vague goals like, "improving market position in all business units," "improving intranet presence through multi-color graphics," or "faster access times." Who is going to monitor these objectives? How can they be monitored at all? It is the measurability of objectives that sets IT stars apart from IT laggards.

Stars formulate tangible objectives with an exact time frame

For companies to access their performance levels, the degree of goal accomplishment must be transparent. At IT stars, IT management and IT professionals are evaluated by their performance level. If you can measure performance, you can afford to set demanding standards. Employees view both ambitious standards and functioning metrics positively; this gives them the opportunity to assess the results of their work and see their role in the organization's success. As a result, top management satisfaction with their IT department, and vice versa, is much higher at IT stars than at IT laggards.

SUMMARY

Top management is a key factor in the success of IT stars. Selected members of management make the vital link between IT strategy and corporate strategy, ensuring a comprehensive perspective. Without the appropriate insight (and go-ahead on resources), companies all too often use IT to improve suboptimal processes, rather than first optimizing processes before selecting new IT. Best-practice companies anchor their IT performance targets to overall business objectives and make sure that delivery against these commitments receives top management attention.

RULE 6

CREATE A CUSTOMER-ORIENTED IT SERVICE NETWORK

At many companies, the term *IT organization* is nothing but a substitute for the conventional EDP department, with all its familiar structural problems (see Exhibit III–4).

On average, only 16 percent of IT professionals are below age 30. Staff have an average of 15 years of professional experience, 11 of which they have spent in the EDP department. They were trained in technologies and systems that have largely become obsolete, and frequently have only limited business management know-how. They have not learned to operate IT systems in decentralized departments or play an active role in cross-functional projects. The result is that many IT staff in the manufacturing sector are not sufficiently equipped to meet the demands of modern IT management.

In the core tasks of IT management, the performance of IT stars is better across the board than that of the other cultures. These strategic and implementation-oriented tasks cover three categories of central importance: IT planning, control, and consulting; management of internal operational functions; and IT outsourcing.

Exhibit III–4
Profile of IT Staff

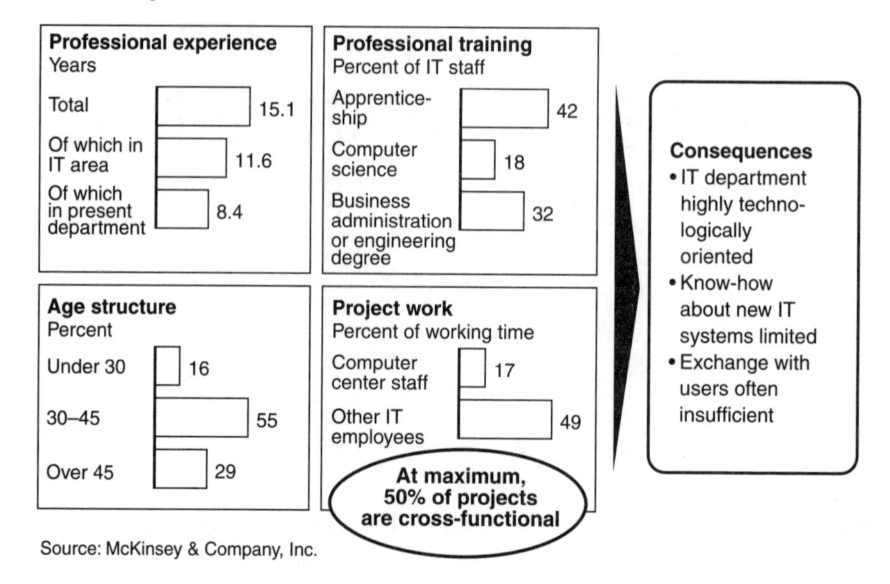

Source: McKinsey & Company, Inc.

IT PLANNING, CONTROL, AND CONSULTING

All companies surveyed consider IT planning, control, and consulting as core competencies and strive to be top performers in these fields. But not all of them meet their high aspirations.

Due to their lack of resources, IT laggards barely manage to look beyond their day-to-day operations. The opposite is true of IT stars. Often they have a dedicated IT management group with the required business process know-how. In the future, they will focus more on the management of planning risks.

LAGGARDS: FREQUENT LACK OF CORE COMPETENCIES

Most of the companies surveyed have not yet succeeded in aligning their IT strategy with their business strategy. In the words of Gartner Group, the IT consultants, “Many board members—mostly members of the techno-phobic caste—do not even see the necessity of reconsidering their IT strategy—assuming they have one at all.” Their IT planning is often a reactive approach based primarily on incremental updating of the status quo. For almost 60 percent of IT laggards, the planning of their future IT development consists merely of extrapolating historical data. Only 25 percent have installed medium-term IT project planning. Laggards make almost no use of planning tools such as project portfolio techniques.

Only 25 percent of laggards have installed medium-term IT project planning

The tasks of IT management at IT laggards are determined primarily by their day-to-day priorities. They have scarcely any defined standards for systems architecture or application systems. Only around one-third use a three- to five-year time horizon for planning (e.g., for hardware, network technologies, and future software). Firefighting often has priority over controlled and systematic IT development.

A key reason for the know-how deficiencies at IT laggards relating to current technologies and applications is that their knowledge development tends to be coincidental rather than systematic. The integration of state-of-the-art IT systems often seems to have low priority. In the planning phase,

BEST-PRACTICE IT ORGANIZATION

MECHANICAL ENGINEERING COMPANY

The IT organization of a leading European mechanical engineering company represents best practice (see Exhibit III–5). IT management and the IT service center are clearly separated. The IT manager is graded on the second management level, and is not an IT expert but a top executive with broad management experience. He is in charge of proposing IT strategy and the IT project portfolio. He works with a steering committee consisting of general management, the division heads, the head of the IT service center, and the head of the user core team. The steering committee decides on the IT strategy and IT project portfolio.

The user core team consists of lead users from all divisions. Its main task is to define user requirements. To do this, lead users rely on their knowledge about business processes and the problems they face in day-to-day operations.

The IT service center is in charge of providing infrastructure and systems operation and employs specialized staff for this. If necessary, cross-functional project teams can also be set up consisting of IT users, internal IT professionals, and perhaps external consultants.

Exhibit III–5
Best Practice for Organization of the IT Function:
A Mechanical Engineering Company

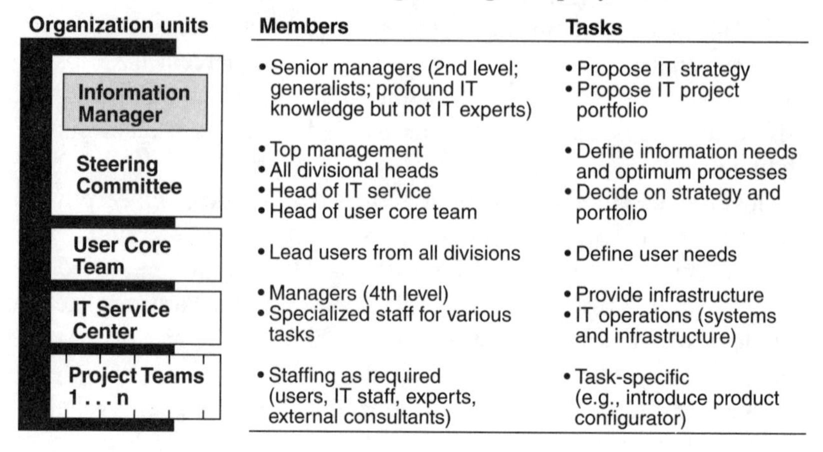

Organization units
Members
Tasks
Information Manager
Steering Committee
• Senior managers (2nd level; generalists; profound IT knowledge but not IT experts)
• Propose IT strategy
• Propose IT project portfolio
• Top management
• All divisional heads
• Head of IT service
• Head of user core team
• Define information needs and optimum processes
• Decide on strategy and portfolio
User Core Team
• Lead users from all divisions
• Define user needs
IT Service Center
• Managers (4th level)
• Specialized staff for various tasks
• Provide infrastructure
• IT operations (systems and infrastructure)
Project Teams 1 . . . n
• Staffing as required (users, IT staff, experts, external consultants)
• Task-specific (e.g., introduce product configurator)

Source: McKinsey & Company, Inc.

the exchange of ideas and experience with IT users is also insufficient. At almost 75 percent of all laggards, there is no systematic scanning of new trends on the IT market. This lack of commitment is reflected right down to the user utilization rate, which is much lower at laggards because users are often only superficially involved in systems selection and implementation.

Many IT laggards have recognized the problem but are unable to solve it alone without fundamental organizational changes. Usually they lack the personnel resources for professional IT management. Additionally, they have significant deficits in their organization structure—for instance, in task definition, delineation of competencies, and performance measurement. "Frequently they do not have the necessary structures," criticizes an IT consultant. "There is no glory involved in being an IT manager at companies like that. That's why you hardly find any good specialists."

BEST PRACTICE: BUILDING COMPETENCE IN A DEDICATED IT MANAGEMENT GROUP

Usually IT stars establish a dedicated IT management group to implement strategic tasks. This area is distinctly demarcated from operational tasks such as IT operations or systems implementation. Its responsibility covers

IT planning and control—above all, the strategy relating to technology and applications—as well as IT-related consulting, including support of business process redesign.

This provides IT stars the necessary organizational structure to focus on proactive, business-process-oriented IT planning. Almost 50 percent of them use the zero-base approach, reassessing projected and current tasks every year, with a planning time horizon of three to five years. This approach ensures controlled IT development in a systematic, bottom-up planning process (see Exhibit III–6).

Lead users are closely involved in the strategic planning process and are usually assigned to this role for a significant share of their time. They have specific know-how about using IT in business processes and user requirements. At more than 50 percent of IT stars, lead users have been nominated in the operating departments.

Over 50 percent of IT stars nominate lead users

IT stars can draw to a large extent on their own know-how about business processes, application planning, technology planning, and consulting. Apart from using this know-how to improve IT performance in their processes, they provide readily available consulting expertise for their

Exhibit III–6
Long-Term Planning for IT Development

IT stars plan their IT development systematically and over the long term

Percent of companies surveyed

IT stars
IT laggards

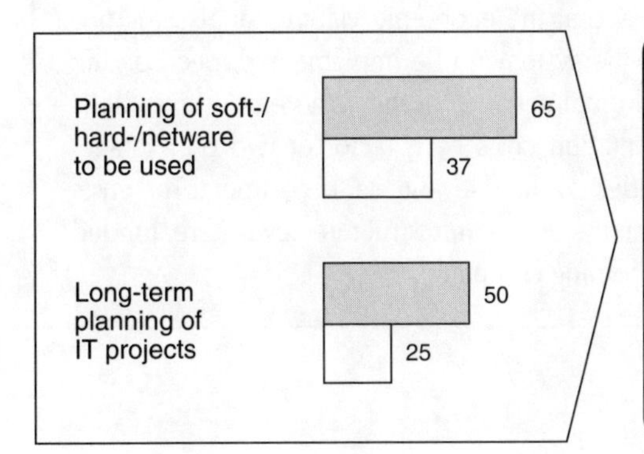

Benefits for IT stars
- Clear strategies for architecture and software standards
- Regular adjustment of IT project portfolio
- Scenarios for resource requirements
- Controlled, systematic IT development instead of firefighting

Source: McKinsey & Company, Inc.

PLANNING THE IT PROJECT PORTFOLIO

At a German company, all IT projects are classified according to four basic categories, each with a different planning process:

- **IT projects for fulfilling legal regulations**—for example, modification of the IT system for accounting due to altered accounting requirements
- **IT projects required for operational reasons**—for example, modification of key operational indicators in the executive information system after introducing teamwork in a factory
- **Rationalization projects**—projects with a medium-term impact on costs, time, and/or quality of business processes, such as introduction of an EDI interface to customers
- **Innovation projects**—projects that support strategic corporate goals but are hard to quantify, such as the introduction of an intranet as a communication technology between different plants

The planning goal for the first two project categories is to ensure the fastest, most cost-effective, and smoothest implementation. For rationalization and innovation projects, the key issue is to find the right IT system.

For the first two project categories, only a rough assessment is made, and then an implementation plan is prepared.

With rationalization and innovation projects, however, detailed investment plans are made. For major projects, the IT manager prepares an appropriate management report in cooperation with future users. It normally consists of a business case, investment proposal, economic viability analysis, payback period, and risk assessment. For IT innovation projects, a benefit analysis is made instead of the payback period calculation.

It is crucial for the company that the economic viability analysis is oriented to the life-cycle costs of the systems to be implemented, because the recurring costs for operation and maintenance of the IT systems at five years' operating life exceed implementation costs by a factor of two on average. Rationalization projects are funded by the relevant user department. The costs for innovation projects—especially on an infrastructure level—are funded cross-functionally (e.g., by the holding company).

users. They receive positive ratings for this from their IT users and lie significantly ahead of IT laggards, whose IT users rate them as mediocre to poor.

Additionally, IT stars rigorously use external technology expertise, partly to gain access to new know-how. They use external consultants much more intensively than laggards, particularly to develop IT concepts, ensuring independent verification and leading-edge IT development proposals.

A special strength of IT stars is their systematic IT market scanning to identify IT innovations early. Almost 90 percent of IT stars evaluate new information technologies and software regularly, involving experts from vendors in the process. More than one-third of IT stars even assign IT professionals specifically to do market research.

One financial service provider is a good example of how to organize this market scanning. This company operates a high-tech studio with an IT reconnaissance team consisting of young computer specialists, physicists, and industrial engineers. They focus on IT developments that hold high promise for the future, but are hard to communicate to their top management. Among other things, they examine future applications of chip cards or electronic commerce and try to identify opportunities for gaining a competitive edge.

Organizational measures of the type described are complemented by sophisticated IT planning tools, such as an application and technology portfolio, resource planning, indicator systems, and scenario techniques. These tools are widely used at IT stars to define priorities for application systems and hardware development. The application systems portfolio technique is used, for instance, to identify suitable software and assess proposed investments. Priorities are defined by criteria for corporate success—for example, what business processes are most relevant, how important the application is for the process, or the operational urgency of a new systems solution. Using these planning tools enables companies to scuttle uneconomical IT projects early and eliminate bottlenecks in the systems development pipeline.

OUTLOOK: BETTER RISK PLANNING

The complexity of IT management is escalating. Technology itself is transforming at breakneck speed, triggering intricate demand dynamics. Organizations are progressively decentralizing into smaller units whose only link is often IT. Companies need to minimize the risks of this com-

IT-DRIVEN INNOVATIONS IN BUSINESS PROCESSES

U.S. AUTOMOTIVE MANUFACTURER

A U.S. automotive manufacturer is currently undertaking vast efforts to improve its business processes with IT projects. Its goal is to focus IT use on innovative applications. Projects are classified in four categories:

- **IT applications with high future potential**—projects that are not yet ready for use in day-to-day business but are in a development stage (e.g., virtual reality for the sales department)
- **Strategic applications**—applications for gaining competitive advantage (e.g., rapid prototyping, digital mock-up)
- **Core applications**—necessary for core business processes (e.g., PPS systems)
- **Supporting applications**—nice-to-have features (e.g., electronic slide show for presentations)

As a rule of thumb, the company tries to allocate about 20 percent of IT staff to the first category, 40 percent to the second, 30 percent to the third, and a maximum of 10 percent to the last category.

This is because the company believes only the first two project categories create value-added, and therefore minimizes expenditure on the last two.

plexity by shifting their IT planning to a new level of precision and speed. To ensure a 360-degree perspective, they should involve both users and external consultants, employing sophisticated IT-based planning tools (e.g., scenario techniques) as appropriate. A carefully structured process can identify planning risks and develop effective countermeasures. Best-practice organizations are already developing their IT planning process in this direction.

IT OPERATIONS, USER INTEGRATION, AND IT CONTROLLING

In addition to planning and control functions, IT management has to ensure the fulfillment of numerous operational IT tasks: data processing

BEST-PRACTICE IT PLANNING

SOFTWARE COMPANY

A leading international software company is already using state-of-the-art planning techniques. With the migration from mainframe to client/server technology, the firm made substantial changes to its European IT organization. In the course of this reorganization, a project team for IT planning was established, with parity representation of top management and divisions. The goal: to tailor IT planning to the business objectives and requirements of internal customers.

To prepare the planning process, decentralized internal IT business consulting teams derive requirements from their own expertise and discussions with the divisions, and then develop project proposals. With the assistance of relevant centralized and decentralized functions, the IT planning team decides which of these projects to pursue. All analyses from the divisions are structured and consolidated to identify interrelationships and select important projects. The prioritization criteria are based on group objectives, and link strategic importance for the entire company to operational urgency for the different divisions. Investment appraisals and benefit analyses complete this portfolio approach.

The job of the decentralized IT business consulting teams is to prepare data to meet decision criteria in a prestructured and standardized analysis. Nevertheless, not all criteria can be measured objectively. Their importance is partly determined as a result of discussions. Final assessment and selection are made by the project team according to the 80:20 rule. (This rule is the Pareto principle: concentrating on the "right" 20 percent of the work produces 80 percent of the results.)

In 1995 the company used this approach to analyze about 350 project proposals, adding up to a total of 30,000 man-days and with a planning horizon of 18 months. The selection methodology successfully prioritized the 20 most important projects.

center and server services, user help desks, network management, and the maintenance of application systems. It is also responsible for projects such as the development of proprietary software and implementation of standard software. Today most of these services are provided by IT organizations themselves, and the plans of many of the companies we surveyed indicate this will not change substantially in the future.

Our findings show that IT laggards have got caught in a vicious cycle of intransparent cost performance data, insufficient tracking of efficiency deficits in IT operations, and ineffectiveness of their own action planning.

IT stars are better off. Distributed, networked service centers strengthen the empowerment of IT users and allow professional controlling of IT costs and performance. In the future, optimization efforts are likely to focus on further involvement of IT users and greater modularization of IT services.

LAGGARDS: CAUGHT IN A VICIOUS CYCLE

The gap between stars and laggards is particularly high in IT operations. But these deficits often remain hidden due to insufficient communication and inappropriate controlling structures.

Our performance ratio analysis shows clear results (see Exhibit III–7). IT laggards' IT costs are 2.4 percent of sales, while 2.3 percent of their employees are IT staff. IT projects at laggards do not meet their budgets; the budget overrun of project costs averages 35 percent. Deviations from plan

Laggards' project budget overrun averages 35 percent

Exhibit III–7
Comparison of IT Efficiency

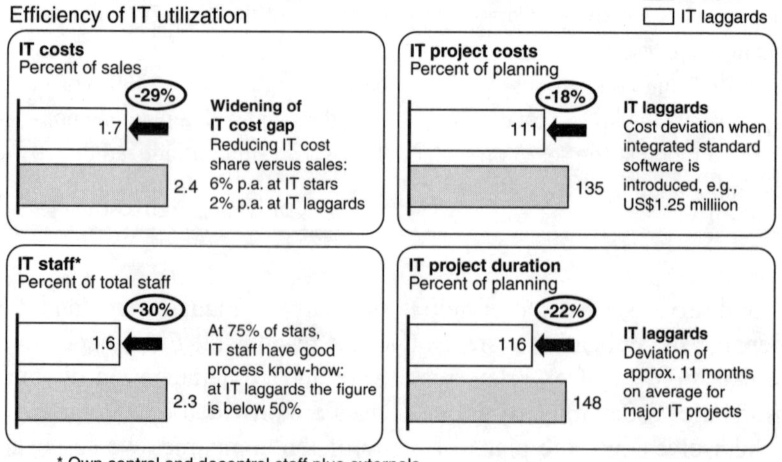

on project duration are even larger (around 50 percent). Each of these performance indicators is 20 to 30 percent lower than those of IT stars.

Often the know-how of the head of EDP and IT staff at laggards is limited to company-specific technologies and systems. This combined with the high dependence of IT user departments on IT support services can have severe repercussions:

- **Insufficient consideration of user requirements**
  Even when new IT systems are implemented, the traditional technology know-how of these IT departments dominates. The new IT systems do not adequately take into account the requirements of user departments because they were specified according to narrow IT criteria. IT users are also not sufficiently involved in systems development and systems implementation. Only 26 percent of IT laggards—in comparison to 53 percent of IT stars—define interfaces between data processing and operational departments explicitly. It is no wonder that at laggards, users rate the effectiveness of IT support as mediocre at best. On average IT laggards are rated only 43 out of 100 points.
- **Missing link between IT systems development and business process redesign**
  Software development and business process reengineering are largely pursued as separate activities. The IT staff at laggards are rarely able to provide consulting support for process reengineering, and are only marginally involved in redesigning business processes, if at all. Our analysis of big reengineering projects at IT laggards showed that 25 percent of these projects were carried out without any participation of IT staff. In half of the projects, just one IT specialist was involved as a member of the project team.
- **Lack of IT cost and performance transparency**
  Only 7 percent of IT laggards—in comparison to 53 percent of IT stars—have a detailed and reproducible IT cost structure analysis available. Forty-five percent said they could give only a rough estimate of their IT costs; the others would need to undertake laborious research (see Exhibit III–8).

  Only 7 percent of laggards can provide detailed, reproducible IT cost structure analyses

Only the direct costs of IT departments are recorded, if at all. This means that the IT costs incurred in the business departments remain unconsidered. Often only hardware costs and major software are counted as an investment. Project proposals for application systems cover only procurement costs—neglecting running costs—and often do not include any

Exhibit III–8
Transparency of IT Costs

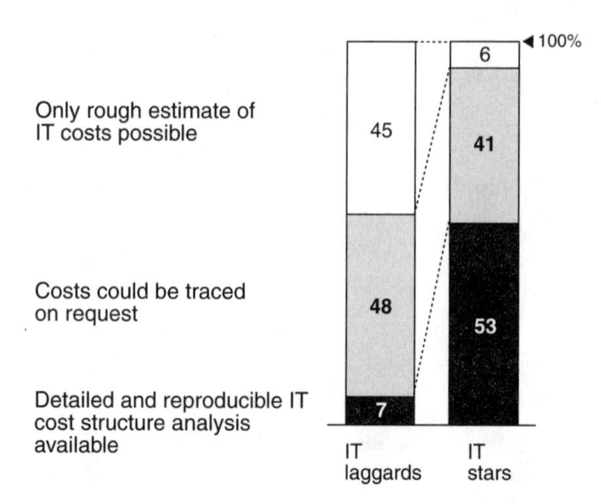

Source: McKinsey & Company, Inc.

measurable performance indicators. This lack of quantifiable criteria also prevents follow-up controlling of project benefits.

IT laggards have almost no possibility of overcoming these difficulties. Their lack of tools for management and IT controlling prevents them from even properly identifying the performance gap. And this in turn prevents the laggards from realizing the need for action. This means they cannot derive or initiate appropriate countermeasures.

BEST PRACTICE: DISTRIBUTED CUSTOMER-ORIENTED SERVICE NETWORKS

IT stars have started to establish internal IT service centers for users. Driven by new technological possibilities and changing business requirements, central EDP facilities are increasingly being replaced by distributed and networked IT units. These are cross-functional service units offering support to several user groups or, alternatively, are decentralized and located in the relevant user department.

The new service centers focus on specific, clearly defined services, such as help desks for IT users, network management, and desktop maintenance. Frequently they work as cost centers. As we discovered, several IT service centers have achieved top performance in comparison to external providers of those services. To ensure this high standard, IT stars regularly perform benchmark tests. In addition, they invite competitive bidding, where internal service centers compete with external suppliers.

What are the specific strengths of IT stars in IT services? Our research revealed five drivers of performance.

INTERNAL CUSTOMER-SUPPLIER RELATIONSHIPS

IT stars develop true customer-supplier relationships between IT management and IT users. Essentially these are based on two conditions. First is a clear agreement on the service level. This means that the price-performance ratio must specified and should be measurable, as the example in Exhibit III–9 illustrates.

Clearly defined quotations or performance specifications include a description of the service packages (hardware, software, or service), competitive prices, and user-oriented (instead of technology-oriented) results. This allows users to cover their individual needs based on a cost-benefit perspective. Moreover, IT management as a supplier must try to find the most efficient solution in competition with external providers.

Exhibit III–9
Price-Performance Comparison: Office Automation

IT stars establish a true customer-supplier relationship between IT management and users

Specification of supplies from IT department

Hardware

- Desktop (Pentium 133, 48 MB memory, 1.6 GB hard disk, 256 KB cache, CD-ROM, 17" SVGA screen, . . .)
- Laptop (Pentium 120, 24 MB memory, 1 GB hard disk, CD-ROM, 12.1" color display, . . .)

Software

- Operating system
  -NT 4.x
  -Win 95
- Software
  - MS Office
  - Antivirus
- Special software (optional)
  - MS Project
  - PageMaker
  - . . .

Service levels

- Installation free of charge
- Exchange of defective units in 24 h (80% in 4 h)
- Support within 1.5 h (80% in 45 min.)
- New machines supplied within 1 week
- On migration, maintenance of two release versions over 1 year

Prices invoiced

- Desktop (HW plus SW): $175 per month
- Laptop (HW plus SW): $275 per month
- Operations (maintenance, help desk, server, printer, data security): $125 per month and client
- Network (connection, e-mail, Internet): $75 per month and client

Source: McKinsey & Company, Inc.

CUSTOMER FEEDBACK SYSTEM

PROCESS INDUSTRY

A company in the process industry set up a special feedback system to improve internal customer satisfaction. Unlike conventional measurement systems, it does not focus on quantifiable cost-benefit criteria, but on performance as perceived qualitatively by internal customers. The goal is to improve acceptance of the IT department. The customer feedback system consists of the following elements:

- **Customer surveys and interviews**
- **Questionnaires for feedback on software projects**
- **After-sales interviews**
- **Complaint management**
- **Problem solving via a help desk**
- **Service letterbox**

This system is designed to focus the attitudes and behavior of IT staff on improving service quality.

The company soon found that internal customers used these instruments only if they could expect *immediate* help for their problems. The acceptance of the system depends on the approach and the behavior of the IT professionals. After this system had been in place for nine months, clear improvements of IT services were apparent. The company aims to improve internal customer satisfaction by 20 percent.

During systems implementation, the quality of this customer-supplier relationship is tested every day. Systematic follow-up can show whether the internal IT users are satisfied. This work helps to identify the changing needs of IT users or deficits in service early on, and steps can be taken to improve customer satisfaction.

INTEGRATION OF IT USERS

IT stars strive for the close involvement and participation of IT users not only in developing IT strategy but also in IT operation and concrete project work. With this attitude, IT stars are far ahead of IT laggards. They successfully pursue the policy of involving lead users from the operating departments. On average, lead users are assigned 20 percent of their time to defining systems requirements, identifying potential for improvement, cooperating in system tests, or even acting as first-level support in the operating department. From the users' perspective, this is a very successful approach. IT stars give a 60 percent higher rating to the effectiveness of lead users than IT laggards do.

SUPERIOR IT PROJECT MANAGEMENT SKILLS

The project management skills of IT laggards are evaluated as mediocre at best, while the consulting and application experts of IT stars are generally rated "good" to "very good." Stars have a strongly developed know-how regarding the redesign of business processes and can access any information or skills they need externally. At IT stars, the implementation of an IT system is based on a detailed business process requirements analysis that is competently supported by IT management. We found that 40 percent of redesign projects were initiated by IT management; in about one-third, IT management provides the team leader, and in two-thirds it has a further full-time member on the project team.

CONTINUOUS IMPROVEMENT OF SYSTEMS PERFORMANCE

The goal of IT stars is to continuously improve the performance of their systems. More than half of these stars obtain clear quantification of their performance levels through comprehensive and regular reviews of all important IT application systems (see Exhibit III–10). In these reviews they focus on criteria such as system utilization rates, user satisfaction, efficient systems use, and compatibility of systems with the corporate environment. IT stars use these results to make the use of IT systems efficient and identify possibilities for improvement.

PROFESSIONAL IT COST AND PERFORMANCE CONTROLLING

IT stars use a sophisticated cost controlling and performance measurement system to evaluate and monitor the impact of IT on business processes. Because IT costs and performance are transparent, external service providers have almost no chance of taking them in with exaggerated specifications or excessive quotations. Moreover, the controlling tools allow efficient IT budget planning that is derived direct from the strategic IT plan. IT stars try to avoid large fluctuations in the IT budget through continuous upgrading of IT systems, for example, and flexibility when unexpected developments arise in day-to-day business.

Exhibit III–10
Frequency and Scope of IT Systems Reviews

At stars, IT staff increase IT performance through proactive advice to users

Percent of companies surveyed

IT stars
IT laggards

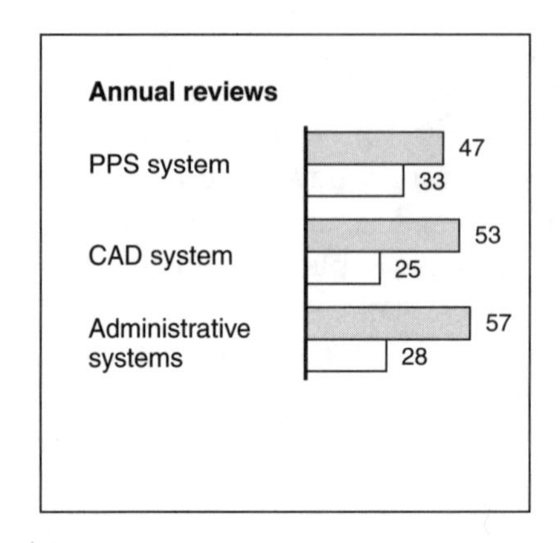

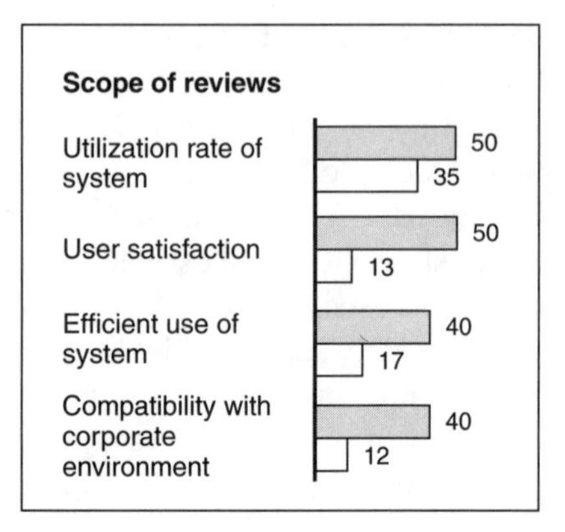

Source: McKinsey & Company, Inc.

OUTLOOK: INTEGRATION OF CUSTOMERS AND FURTHER MODULARIZATION OF IT SERVICES

Undoubtedly there is a trend toward increasing the integration of external end users into the business processes of manufacturers. This means that IT management cannot confine itself in the future to supporting internal users and suppliers; increasingly it must create interfaces to customers as users of company-specific IT. First examples of IT use at the customer interface are virtual product demonstrations based on graphical animation and visualization (see Rule 1 and Part IV).

Apart from such technology-driven innovations, IT management must contribute to supply chain management—integrative management of the sequential flow of logistical, conversion, and service activities from vendors to ultimate consumers—as described in Part IV. Increasingly IT will make it possible to optimize individual steps of the supply chain and thus open up synergies for the entire business process. It should be the task of IT management to press ahead with the modularization of IT services in such a "fractal" company. The operational IT center will need to combine

STANDARDIZED ASSESSMENT OF IT INVESTMENTS

AUTOMOTIVE MANUFACTURER

An automotive manufacturer found that the conventional assessment of IT projects on the basis of return-on-investment calculations was misleading. Major reasons were that directly allocable costs of IT systems implementation, such as training and operating costs, were regularly estimated inaccurately, and indirect or hidden costs (e.g., reduced productivity in the learning phase) were largely neglected.

The company developed a standardized assessment for recording the costs, time, and quality effects of an IT investment precisely. Cross-functional teams representing the IT organization, controlling, and user departments are in charge of the analysis. These teams have access to a special database that contains the data for all IT projects. This ensures that the team's cost-benefit analyses are consistent and comprehensive. The continuously updated project database results in an "intelligent" system where new data are regularly available and all process improvements and new team methods are gathered systematically.

local, decentralized presence with overarching integration and knowledge access.

IT OUTSOURCING

Outsourcing has become accepted business practice and is frequently the obvious approach to optimize the use of IT in organizations. Today the range of outsourced services is very broad, covering all the IT operations just discussed: IT project services (e.g., application development and systems implementation) and services closely related to IT (e.g., consulting services, engineering services, and operation of IT-intensive business processes such as bookkeeping and settlement of accounts).

The four IT cultures all use outsourcing intensively. About three-quarters of the companies surveyed obtain at least part of their IT services from third parties. But this does not mean that IT stars and laggards have equal skills where make-or-buy decisions are concerned. On the contrary, there are vast differences. And in view of the further differentiation of outsourcing services ahead, we expect this gap to widen.

LAGGARDS: SERIOUS PLANNING AND IMPLEMENTATION DEFICIENCIES

In IT outsourcing, laggards prefer to focus on IT operations and project services (see Exhibit III–11)—resource-intensive data center tasks and the development of proprietary applications. Their main objective is to optimize IT costs. Criteria like improving service quality, access to new technologies, or fast know-how transfer generally play a secondary role.

Laggards focus their outsourcing on IT operations and project services

In outsourcing they suffer from a series of characteristic management deficits that affect all stages of the outsourcing process and influence both the planning and execution of outsourcing projects. The consequences are often that IT laggards lose control over outsourcing and become dependent on external providers.

The main problems of IT laggards in IT outsourcing are excessively narrow targets, insufficient control of the outsourcing process, and vague contractual arrangements.

Exhibit III–11
IT Outsourcing Practice of IT Laggards

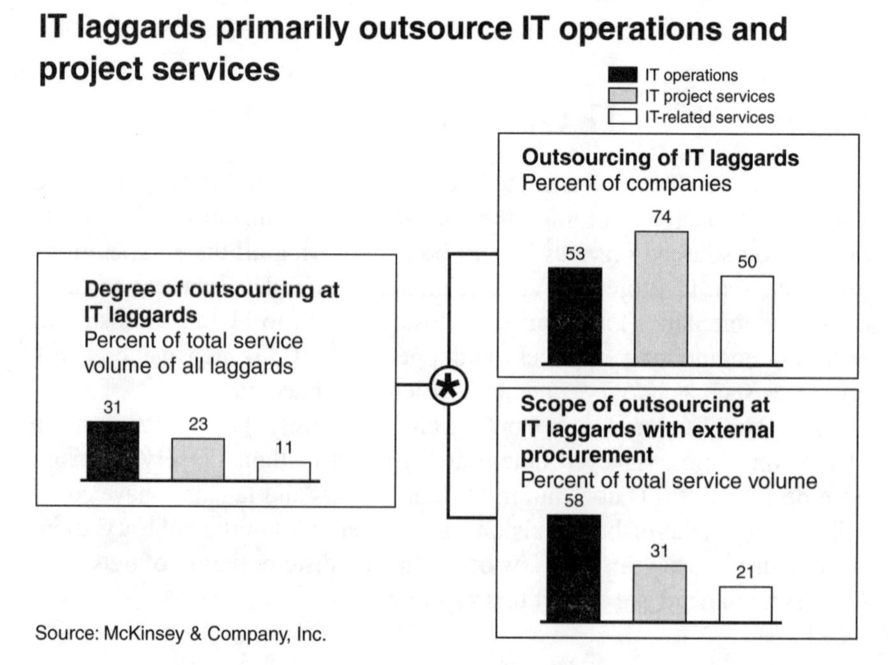

EXCESSIVELY NARROW TARGETS

Usually IT laggards do not aim high enough because they are guided not by best practice but by their own uncompetitive performance level. Moreover, their view of potential benefits is distorted by their focus on supposedly objective efficiency targets, ignoring potential add-on benefits. One interviewee stated, "We need a simulation calculation for each new tool configuration. We aren't concerned with tool optimization but only whether we will get the safety certification. The only thing that counts is who can provide us with this service at the lowest price."

INSUFFICIENT CONTROL OF THE OUTSOURCING PROCESS

At the beginning of an outsourcing relationship IT laggards often have high hopes for the "strategic partnership"—hopes that can wear thin very quickly. The lack of systematic controlling turns out to be a serious constraint on the relationship. Typical problem areas for IT laggards are lack of continuous planning and control of user requirements, no regular comparisons of target versus actual cost-benefit ratios, and no specific controlling of outsourced development projects. In many cases, the former EDP manager is assigned to the role of relationship management without the required capabilities or tools.

It is no wonder that reports about failed outsourcing partnerships have been on the rise, and that IT managers are increasingly criticizing the unrealistic outsourcing expectations of their own companies.

VAGUE CONTRACTUAL AGREEMENTS

Frequently companies make general contractor arrangements without detailed service specifications. They expect greater efficiency from the long contractual period and fixed budgeting, but do not explicitly agree to any additional planning or negotiation cycles, although they know market changes will occur. Direct consequences are high risks such as shifts in demand versus long-term planning, strong lock-in of the service provider, or high dependence on a single provider, all of which pose great dangers for customer and outsourcer alike.

BEST PRACTICE: FOCUS ON IMPROVEMENT OF EFFECTIVENESS AND PROCESS CONTROL

IT stars use a very different approach. Usually they already have an efficient IT management and a clear idea, derived from benchmarking of their own IT organization's performance. In their make-or-buy considerations

IT OUTSOURCING AS A BUSINESS RISK

UK-BASED CONGLOMERATE

Recently an outsourcing contract with a volume of around $570 million between a UK-based conglomerate of various consumer goods manufacturers and a leading global outsourcing provider failed after 2 years, even though a period of 10 years had been agreed on. In the first phase, the IT organization was completely restructured and transferred to a central organization with 200 employees, and operation of the total IT infrastructure was taken over by the outsourcing provider. Then unexpected market changes caused the group to redefine its business strategy. This resulted in massive divestment and work-force reductions, so the basis for the partnership changed dramatically. Many problems were caused by the inflexible contractual arrangement, centralization designed for growth, and the vague performance specifications, all of which now need resolving in an elaborate and tangled process.

Exhibit III–12
Objectives of Outsourcing for IT Stars

IT stars concentrate on improving their IT effectiveness through outsourcing

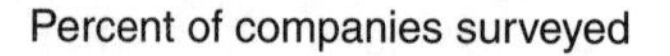
Percent of companies surveyed

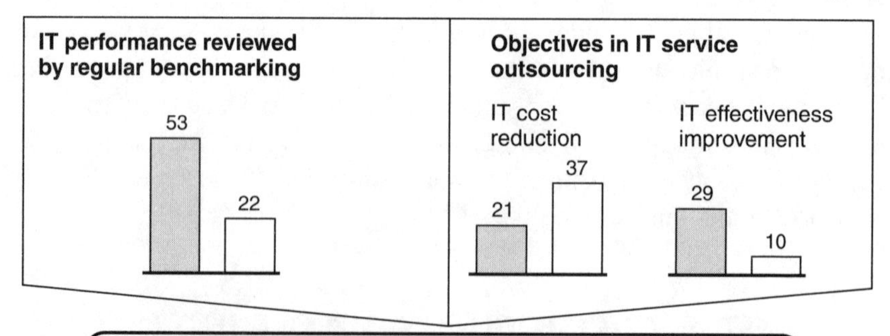

- High in-house efficiency of IT stars puts them in a good competitive position versus external outsourcing providers
- IT stars concentrate on IT effectiveness improvements
- IT laggards focus on cost reduction through outsourcing

Source: McKinsey & Company, Inc.

they therefore focus primarily on optimizing IT effectiveness (see Exhibit III–12).

The decision to make higher effectiveness the top priority of IT outsourcing means that the approach of IT stars is quite different from that of IT laggards (see Exhibit III–13).

IT OPERATIONS ARE RARELY OUTSOURCED

Only about 25 percent of IT stars outsource IT operations at all, and the share of total service volume they outsource is 29 percent in comparison to 58 percent at IT laggards. The reason is that IT stars often have the skills to ensure efficient IT operation in-house. Where the classic outsourcing of EDP centers is concerned, economies of scale are of little relevance because many IT stars are already migrating from mainframe computing to client/server architecture.

Degree of outsourcing at stars is highest for IT-related services

Exhibit III–13
IT Outsourcing Practices of IT Stars

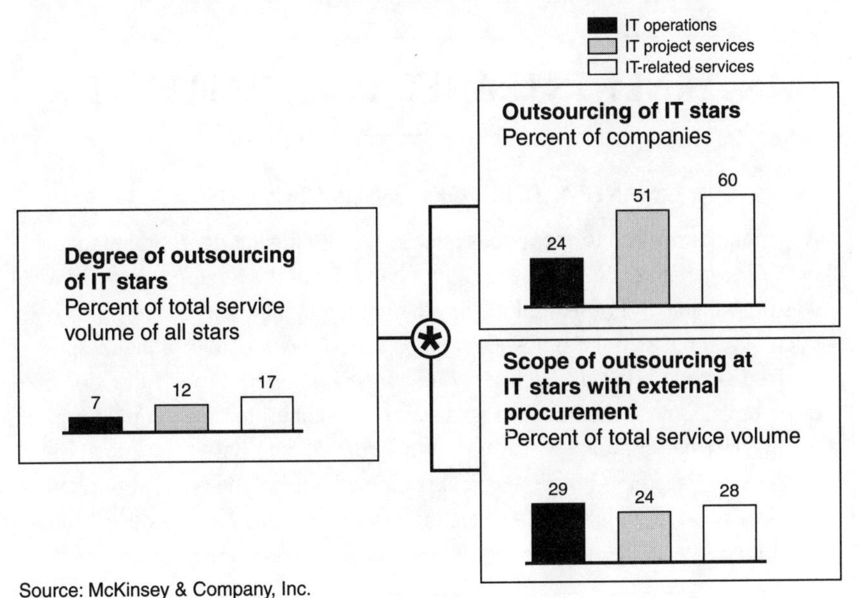

Source: McKinsey & Company, Inc.

PROJECT SERVICES ARE OUTSOURCED SELECTIVELY

Half of IT stars outsource development work. The primary objective is to level out their capacity bottlenecks. They select specific application system development projects and outsource them as a complete package. Outsourcing makes up around 25 percent of stars' total project volume on average.

GROWING OUTSOURCING OF IT-RELATED SERVICES

IT stars are faster to outsource services closely related to IT because by involving external expertise, they seek to ensure fast learning and focused development of know-how within their own organization. The degree of outsourcing at IT stars is 17 percent of their IT-related services, in comparison to 11 percent at IT laggards.

IT stars understand the importance of supplier management as a key factor for success in outsourcing. They concentrate on contract design and partner management when building the partnership.

Their outsourcing contracts are characterized by clear specification of which tasks are to be performed externally and which internally, detailed service level agreements based on user benefit criteria (e.g., response time, systems availability, equipment for client workplaces, and flexibility).

An institutionalized joint planning process between outsourcer and service provider allows the dynamic adjustment of contractual arrangements to market-driven shifts in requirements. Opportunities and risks can be dis-

INTERNET/INTRANET AS A CHALLENGE

MANUFACTURER OF CONSUMER GOODS

A manufacturer of consumer goods intended to use the Internet/intranet more heavily as a marketing and distribution channel, but the required know-how was unavailable in-house, and the life span of Internet/Intranet applications is often only three to four months. The company decided to employ a highly specialized service firm, at which computer specialists work around the clock, using development expertise across several time zones, to develop platforms for the Internet. Simultaneously the firm is developing base web programming/marketing know-how in its organization without taking on new employees. In view of the uncertainty of technology development, it has decided to continue its focus on outsourcing for know-how development in this field.

tributed more evenly between the contracting parties. In some cases, outsourcing contracts no longer have fixed-price agreements but are based on risk and profit sharing. Moreover, companies are increasingly modularizing the IT services they wish to outsource into dedicated bundles and outsourcing them to different providers. For example, after its main contractor agreement expires, an automobile manufacturer plans to enlist other providers for a large share of its outsourcing volume, starting with the development and implementation of an order processing system and desktop management.

These changes make high demands on partner management. Existing tasks are being supplemented by the new function of broker, responsible for internal and external IT services, services integration management, and liaison when multiple providers are used. The management of dynamic partnerships with several providers is replacing the concept of a strategic partnership with one provider. Regular invitations to tender and placement of orders based on clearly defined criteria maintain permanent competitive pressure between service providers and ensure optimal procurement of services.

OUTLOOK: FROM UTILITY MANAGEMENT TO BUSINESS PROCESS OUTSOURCING

IT experts and financial analysts expect the current outsourcing boom to continue. Depending on the type of service, annual growth rates of 10 to 30 percent are forecast. There are likely to be dramatic changes, with the outsourcing practices of IT stars becoming industry standard.

Classic outsourcing fields such as IT operations and software development will increasingly be supplemented by IT utility management, where an external service firm is assigned responsibility for provision, installation, and operation of a company's total IT infrastructure for its complete life cycle. The Gartner Group, for instance, assumes that the focal point of outsourcing contracts will shift in the medium term to utility management.

At the same time, the trend toward modularization and commoditization will grow. Clearly defined service bundles will be offered for sale in competitive bidding. Market prices will develop for these mass services. The approach of the U.S. government could serve as a model in this context. It has developed a framework for the terms of bidding requirements for outsourcing desktop computer support (e.g., operations, maintenance, technology upgrades) to facilitate the selection of suppliers and obtain comparable offers at lower cost.

IT OUTSOURCING AS A DYNAMIC PARTNERSHIP

COMPONENT MANUFACTURER

A global component manufacturer outsourced its total IT system management—IT operations, software implementation, consulting services for IT users, and other areas—to an external service provider on a long-term basis. The agreement was not restricted to commercial applications: technical IT applications (CAD, EDI, Vax cluster) were outsourced too. Only applications related to the company's core competence in product development (e.g., simulation tools, specific stability calculations) remained the responsibility of R&D. The facility management contract included transfer of all IT assets (hardware, software licenses, staff, buildings, etc.) to the service provider, and runs for 10 years.

The key success factor of this partnership will be to maintain an IT management function at the component manufacturer. A department of five employees was established for this with full responsibility for developing IT strategy, doing market research, and sponsoring IT projects. The department is also responsible for know-how transfer from the service provider to the company and control of the external partner's services.

Two contractual components are especially worth mentioning:

- **Flexibility of contractual arrangements**
  Every year some 10 extensions and modifications are agreed on to adjust the contract to current IT requirements.
- **Exact pricing according to services rendered**
  Differentiated prices are agreed on for the various services (e.g., individual assignment of each project, use of EDP facilities based on time slices) to provide the customer with a clear overview.

Price/performance arrangements are renegotiated every year. The IT department of the component manufacturer ensures that prices are in line with market levels via regular benchmarking of other IT service providers. The external service provider's performance is also monitored by key indicators on a monthly basis.

The objectives of IT implementation projects (e.g., worldwide rollout of an integrated standard software) are defined by IT management and are binding for the service provider. The provider is awarded an implementation contract only if it wins in competitive bidding.

The company's IT department meets the provider at a fixed time every week to ensure intensive communication.

The trend toward business process outsourcing (BPO), a subject discussed since the beginning of the 1990s, is not quite so clear. BPO means entirely outsourcing business processes or functions that are closely related to IT to an external service provider, often sharing both risks and profits.

We found that BPO has won only limited acceptance in the manufacturing sector. Interestingly enough, the proponents in this field tend to be laggards, tempted by the prospect of cost advantages. Overall, the companies surveyed do not see a clear trend toward BPO. They do not even anticipate a rise in outsourcing of strategically unimportant business processes in the medium term. This is not the view of market observers, who expect a big growth for BPO. Business processes that would possibly be suitable for BPO are accounting, invoicing, logistics, sales administration, and sales service support. Companies could benefit from gains from specialization of third parties in fields that are not of strategic importance and cannot be covered with their own core competencies. Contracts with clear performance objectives, strict performance measurement, and systematic performance control by a skilled internal controlling team will be crucial for the success of BPO.

With BPO, as with so many other trends in the IT industry, U.S. companies have a leading edge. Our research indicates that in the United States, the outsourcing of functions such as payroll accounting, invoicing, collection, and financial accounting is sometimes twice as high as at European companies (see Exhibit III–14).

SUMMARY

Stars align their IT organizations to optimize IT management along three key dimensions (see Exhibit III–15). First, they leverage their IT planning, control, and consulting skills in an information management group with close links to top management. This group builds extensive business process know-how and directly supports business strategy. Second, they frequently handle IT operations via decentralized, networked IT service centers. Proactive involvement of users and efficient IT cost-benefit controlling are important prerequisites. Third, IT stars are smart about outsourcing. The main driver behind their make-or-buy decisions is not cost reduction but improved effectiveness. Stars also make sure they build skills that are crucial in the implementation phase, such as contract and partner management.

In the future, IT management will increasingly develop into an independent consulting and service competence for users, with its focus on sat-

Exhibit III–14
Business Process Outsourcing

U.S. companies point the way in business process outsourcing of IT-related services

Percent of companies surveyed

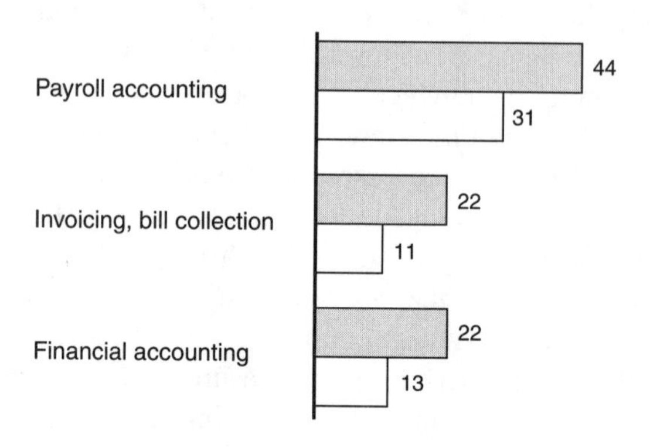

Source: McKinsey & Company, Inc.

Exhibit III–15
Functions of Professional IT Management

Strengthening professional IT management to leverage IT capabilities

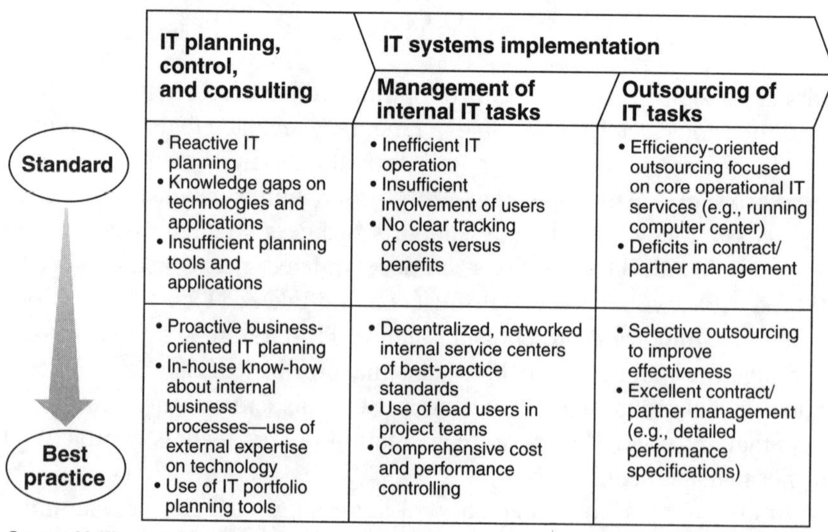

	IT planning, control, and consulting	IT systems implementation: Management of internal IT tasks	IT systems implementation: Outsourcing of IT tasks
Standard	• Reactive IT planning • Knowledge gaps on technologies and applications • Insufficient planning tools and applications	• Inefficient IT operation • Insufficient involvement of users • No clear tracking of costs versus benefits	• Efficiency-oriented outsourcing focused on core operational IT services (e.g., running computer center) • Deficits in contract/ partner management
Best practice	• Proactive business-oriented IT planning • In-house know-how about internal business processes—use of external expertise on technology • Use of IT portfolio planning tools	• Decentralized, networked internal service centers of best-practice standards • Use of lead users in project teams • Comprehensive cost and performance controlling	• Selective outsourcing to improve effectiveness • Excellent contract/ partner management (e.g., detailed performance specifications)

Source: McKinsey & Company, Inc.

isfying internal customers. Its key function will be process redesign and the development of new business concepts rooted in its knowledge of new technologies and their fit with the company's business processes. IT management will also act as a service broker, negotiating, managing, and controlling a web of internal and external service providers.

RULE 7

INTRODUCE INTEGRATED STANDARD SOFTWARE ON A FAST-FOLLOWER BASIS—BUT REDESIGN THE BUSINESS FIRST

Our research indicates that most companies in the manufacturing sector prefer standard software to proprietary solutions. This is especially true of IT stars and sets them apart from the more indifferent attitude of IT laggards (see Exhibit III–16). The most commonly used integrated standard software in the manufacturing sector includes SAP R/3, Baan IV, Oracle Applications, SSA BPCS, J. D. Edwards, and JBA.

In implementing these systems, IT laggards and IT stars alike tend to rely on support from external consultants, but the implementation risks remain high. Despite all efforts to standardize implementation processes and use external consulting services, most companies—above all, IT laggards, but surprisingly many IT stars too—suffer from striking shortcomings in implementing IT systems (see Exhibit III–17).

- **Excessive project length**
  At a company with eight locations and annual sales of $1.5 billion, the implementation of a standard software package has already taken four years, and estimates are that it will take another five to six years.

Exhibit III–16
System Implementation Efforts, by Type of Software

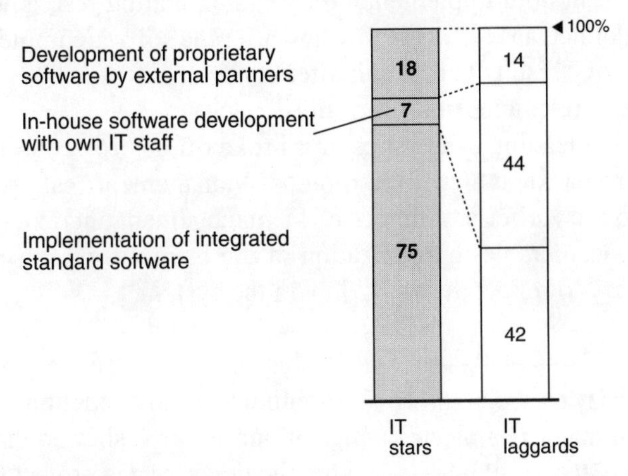

Source: McKinsey & Company, Inc.

Exhibit III–17
IT Project Management Performance: Schedule and Budget

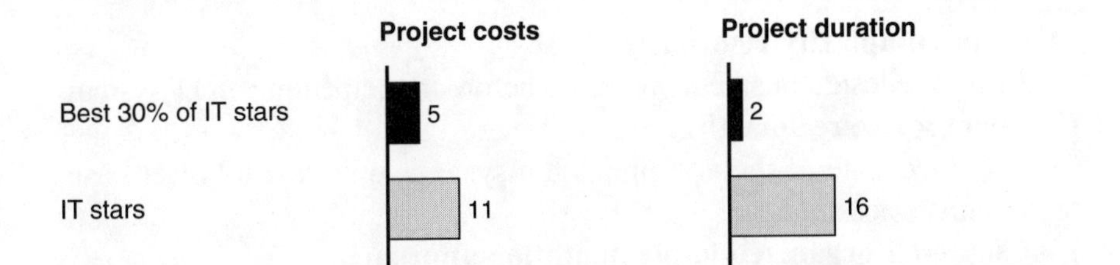

Source: McKinsey & Company, Inc.

- **Big budget overruns without corresponding benefits**
  A European automotive supplier intended to improve performance in order processing by implementing an integrated standard software package (to increase its inventory turnover by 70 percent, for example). It not only failed, but the costs of the implementation project were so high that in both implementation years, operating results were unofficially defined as "results before and after the software implementation" instead of "results before and after tax."
- **Inadequate conceptual planning**
  In 1993, a leading electronics firm broke off its project to implement a software package after nine months. Management realized that, contrary to the earlier assurances of IT management and external consultants, considerable reorganization of the business processes would be necessary to reflect the work flow in the system.

Laggards' delay on project schedule for major IT projects is just under 50 percent on average

This list of negative examples could be continued almost indefinitely. The findings of our research showed that at IT laggards the delay versus project schedule of major IT projects is on average just under 50 percent, or 11 months; the average budget overrun is over a third.

It is easy to imagine the risks this leads to with time-critical projects, such as software adjustment to cope with the millennium bug or the introduction of the euro.

Such deficits in implementation inevitably result in serious budget overruns and extended deadlines. However, IT stars demonstrate that IT systems can be implemented successfully. The best IT stars appear to observe three very simple principles:

- **Limit complexity rigorously.**
  Always redesign business processes before implementing an IT system.
- **Select software smartly.**
  Introduce state-of-the-art application systems on a fast-follower basis wherever possible.
- **Use a well-organized implementation approach.**
  Use microcosms for pilot tests, and strictly follow project management guidelines.

TYPICAL DIFFICULTIES IN IMPLEMENTING IT PROJECTS: FOUR PERSPECTIVES

- *CEO:* "The IT consultants who have been involved in our project to implement the integrated standard software package proved to be absolutely unqualified. They were unable to understand our very specific business processes and to depict these in the software, and they were not sufficiently familiar with all modules of the integrated package to assess the effects that table adjustments in one part of the software might have on other parts of the program."
- *IT manager:* "Users and management left my team and me in the lurch. After deciding to implement the standard software, top management withdrew totally, and the division managers did not deliver the manpower we had been promised from the user departments."
- *IT user:* "Our IT manager wanted to carve his name in stone with the new order processing system. He sits in his ivory tower and—without asking us as users—has bought the latest and most expensive software. Unfortunately, the manufacturing control system isn't compatible with our process."
- *IT consultant:* "The company called us in much too late. We had to debug the most awful blunders to get the project back on track. The CEO had approved this project too quickly; there was neither an analysis of the status quo nor a target concept. The IT manager had no clear objectives, and the users didn't know what their processes should look like in detail."

REDESIGN BUSINESS PROCESSES FIRST

Contrary to common belief, the use of integrated standard software is no guarantee of success; after all, well-known standard software, such as SAP R/2 and R/3, Baan IV, and Oracle Applications, is used by all IT cultures. Analysis of the matrix in Exhibit III–18 reveals that on average, 85 percent of companies with high IT effectiveness use integrated standard software, but only somewhat over 60 percent of companies with low IT effectiveness. This reflects the fact that integrated standard software offers sufficient functionality for most applications and, in combination with state-of-the-art hardware configuration, ensures good availability.

Using integrated standard software in itself does not make a company an IT star. The essential first step is to analyze the business process that

Exhibit III–18
Use of Integrated Standard Software by Type of IT culture

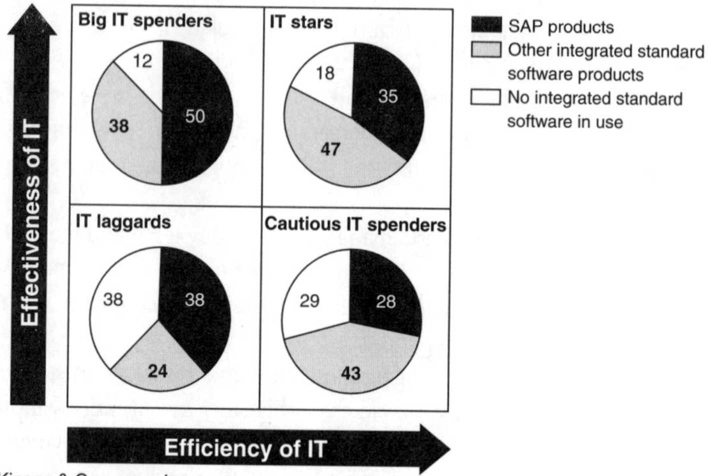

Source: McKinsey & Company, Inc.

you wish to support, only afterward selecting the most suitable software solution. (We will discuss the selection of the "right" standard software later in this chapter.)

Instead of examining the real cause of operational downtime or inefficiencies in business processes, frequently the solution is hastily sought in a new software. The following is a typical example: Although stocks of semifinished and finished goods are increasing in plant X, customers complain about delayed deliveries. The plant manager blames the production planner, who sees the problem in insufficient functionality of the PPS system. The IT manager begins to praise the innovative shop floor technology of an up-and-coming software vendor. A dispute between the production planner and the IT manager follows. After a few minutes of technological jargon, the plant manager is lost. In the end, following the recommendation of the IT department, they agree on the smallest common denominator: implementation of the new vendor's shop floor control system to try to get to grips with the business process.

Integrated standard software in itself does not make a company an IT star

Why, when problems like this arise, isn't the business process the first to be analyzed? A number of different factors are at play. Fear of organizational change is one; another is the communication barrier between users and the IT department. The IT staff's fixation on technology and lack of familiarity with day-to-day routines in the operational departments are further obstacles. A meaningful discussion on how to improve the business process is impossible as long as these fundamental problems rooted in the corporate culture remain unsolved.

At this point, top management must step in to ensure that a scrutiny of the underlying business processes precedes the implementation of a new IT system. This is vital because the decision to implement a new system has a very long-term impact:

Average Operating Life (in years)	
Database systems	10 to 15
Telecommunication systems	10 to 12
Operating systems	8 to 10
Integrated standard software packages	6 to 8

Even if the current work flow seems to be satisfactory, it is important to consider carefully whether the existing business processes can ensure competitiveness in the medium and long term in the face of market dynamics.

It is no coincidence that more than 80 percent of IT stars implement IT systems only in the context of business process design, whereas almost half of IT laggards believe they can do without such a process adjustment (see Exhibit III–19). At best, IT laggards create better support for suboptimal processes. But all too often "the chaos of our order processing is reflected in the software," as one frustrated CEO frankly admitted.

Once a company decides in favor of redesign, the next question is, First process redesign and then systems implementation, or both simultaneously? Here, too, the answer of the best companies is clear: The sequential approach is the most suitable to prevent complexity from getting out of hand and make sure the full potential for process redesign is utilized.

The sequential approach is best

- **Minimize complexity**
  Our research indicates that on average, the sequential approach reduces implementation time by some 50 percent and total project expenditure by more than 60 percent. Above all, it reduces the risk of failure.

  The parallel approach was often used in the past, particularly for urgent projects. The most common argument was, "We can't afford to redesign the process first. There isn't time." Usually the results were disillusioning. Many companies could not make progress because too

Exhibit III–19
Strategies for Implementation of Integrated Standard Software in Order Processing

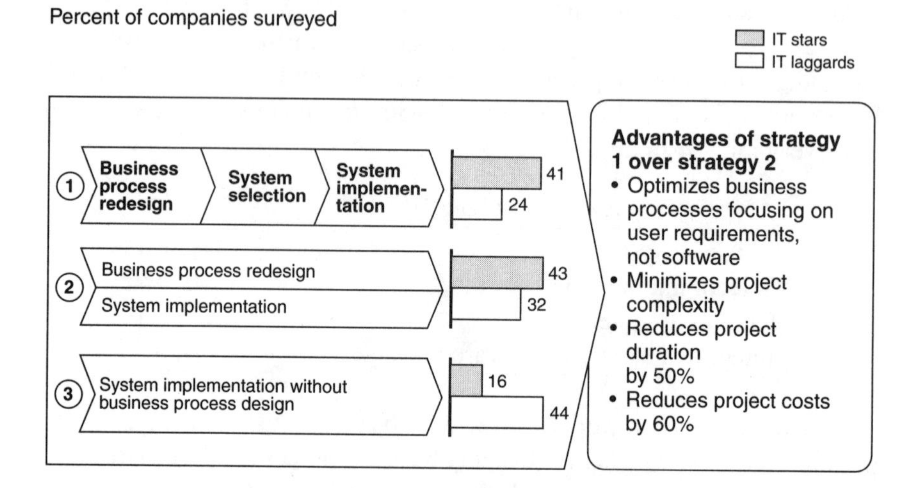

Source: McKinsey & Company, Inc.

many activities had to be performed simultaneously, and they had to call off the project prematurely. Yet the work still had to be completed somehow. Either the process redesign was so rudimentary that it did not deserve the name, or the company was forced to use external resources to make headway with the work—and so project costs exploded.

- **Exploit the potential for optimization**
  In addition to careful analysis of the status quo and benchmarking for concept development, IT stars understand the importance of clearly defined implementation requirements.

  To analyze the work flow and develop a redesign concept, IT stars tap the expertise of external strategy consultants. Alternatively, they may use process consultants, either from their own consulting staff or from a subsidiary that has already carried out similar projects. Their goal is to access the innovative process know-how of these experts and to have their concepts checked by an independent third party.

  For IT stars, the result of the design process is a blueprint of the new work flow with concrete implementation requirements regarding effectiveness and efficiency. How much can stocks be reduced through a better inventory management system? How much should delivery reliability be improved by introducing an integrated supply chain management system? What percentage of reused parts in new designs should be achieved using a comprehensive PDM system?

The most suitable standard software is selected based on these target concepts. Then business processes are fine-tuned to the functional requirements of the best software that is available. Here, too, it is smart IT implementation management that makes the difference for IT stars.

From the beginning, the IT management is involved in redesigning business processes so that its staff can apply their knowledge about the pros and cons of available software at an early stage. They not only examine the constraints of individual systems (e.g., "System X doesn't support bottleneck planning") but also look at their potential for innovative solutions. One participant in our survey told us his attention was drawn to a key opportunity to improve the sales work flow only after comparing various innovative software options. Confronted with the problem of incomplete and inconsistent order data, the IT manager came up with the idea of using the product configurator module that was part of a standard software package. This allowed the entire order process, which had always been extremely trouble prone, to be replaced by a faster, foolproof process. Now sales representatives feed order data directly into the PPS system when they visit a customer.

Limiting process redesign to the development of a rough blueprint for the new work flow allows leeway for software selection and detailed adjustment of individual operations. The production manager of a leading European company confirmed: "Analyzing various software packages greatly helped us to optimize process details. For several problems, solutions we never would have thought of ourselves were served to us on a silver platter."

Will new orgware tools such as business engineering workbench (BEW) and dynamic enterprise modeling (DEM) completely revolutionize the implementation process for integrated standard software, as software vendors have announced? Will they make sequential implementation obsolete? The answer is no. These tools offer only limited support for optimizing conceptual work in the blueprint phase of business process design. They accelerate configuration of the standard software on a microprocess level and enable consistent documentation. Although experts estimate that they can cut implementation costs by 40 to 60 percent, these tools cannot accomplish the crucial conceptual work of business process design.

Orgware tools cannot accomplish the crucial conceptual work of business process design

Only microprocess modules in libraries of reference models for business processes of different industry sectors could conceivably call the sequential approach into question. In the medium term, software firms and consultants are expected to offer modules of this kind where process know-how crucial for successful implementation is directly embedded in the software. But it will take several years before these orgware tools reach market maturity.

GUIDELINES FOR SYSTEMS IMPLEMENTATION

Usually IT stars implement innovative IT systems much earlier than IT laggards do. This applies not only to the latest standard software; IT stars also use more advanced software engineering tools for developing proprietary software.

This does not mean that IT stars are simply technology enthusiasts who want innovation at any price and snap up every new system as soon as possible. They attempt to draw the right conclusions based on their experience and IT market developments—as far as these can be predicted. Obviously IT stars have also had their problems with proprietary software and the implementation of standard software, particularly when the program was not debugged, or was poorly documented. But unlike laggards, they do not develop skepticism or reluctance toward the introduction of new systems.

Instead, stars see IT implementation as an opportunity to improve the performance of their business processes. This fundamental attitude results in careful and farsighted analysis of the IT market for potential to improve performance via new standard software solutions and software engineering tools. Their first question is always how they can gain competitive advantage through the use of IT. They will only develop proprietary software if it will give them a lead over competitors. Otherwise they use integrated standard software.

USE PROPRIETARY SOFTWARE ONLY IF IT WILL GIVE YOU A GENUINE COMPETITIVE ADVANTAGE

The use of IT in itself rarely gives a company a competitive edge. This is clearly illustrated in the cautious approach of IT stars. In general, they take on a pioneer role in only one IT project in six, and primarily focus on implementing integrated standard software packages. However, if they

Stars take on a pioneer role in only one IT project in six

identify a clear potential for differentiation, they exploit it systematically, developing and implementing new special-purpose applications. And in these cases, the risks and costs of a true pioneering feat pay off.

Adjusting existing proprietary software plays a comparatively minor role at IT stars. Only one-third of their expenditure for development of proprietary software falls into this category, while this figure for IT laggards is about 70 percent (see Exhibit III–20).

Where competitive advantage can be gained depends on the general setting—industry structure, market situation, company situation, and other factors. Yet our research identified characteristic success patterns.

Component suppliers with large batch sizes can gain significant advantage from focusing their IT efforts on product development and optimization of logistics links to their customers. Companies with small batch sizes and many product variants or modifications can differentiate primarily by focusing their IT efforts on intelligent product configurators as well as optimal service, aimed at winning the long-term loyalty of their customers. For example, a manufacturer of high-performance servers offers customers a telediagnosis and remote maintenance system, thus significantly reducing the failure rate and often avoiding expensive on-site repairs.

Exhibit III–20
Structure of Development Costs for Proprietary Software

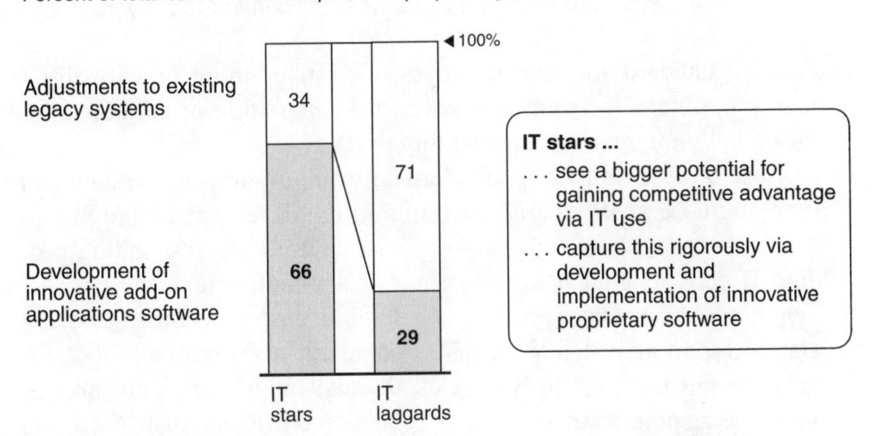

* Services rendered for own account and external procurement of services. (Basis: IT system development projects 1993–1996 with development expenditure of more than one man-year.)

Source: McKinsey & Company, Inc.

The decision to develop cutting-edge proprietary software does not necessarily mean these programs have to be developed in-house. Often it is better to outsource software development. Such cooperation can offer mutual benefits if performance specifications and rules are clearly defined. Most IT stars have created a network of small IT providers that develop special-purpose software for them as strategic partners. Outsourcing makes it possible to vary fixed costs according to actual requirements because staff do not have to be kept available for the development of IT applications. Another even more important advantage is that outsourcing allows companies to tap state-of-the-art know-how and the superior flexibility of specialized service providers. Often these can respond to market and technology developments faster than the company itself, which is usually much larger.

Exclusivity agreements should be the basis for cooperation with service providers to ensure adequate protection of know-how. Rigorous partner and project management is also essential, even for these small service providers.

BEST PRACTICE: INTRODUCE STANDARD SOFTWARE ON A FAST-FOLLOWER BASIS

If competitive advantage cannot be achieved through proprietary IT systems, IT stars go for an integrated standard software package that provides all state-of-the-art functionalities and has already been introduced successfully on the market. This allows them to capture both effectiveness and efficiency gains. Standard software systems offer considerable benefits:

- Usually standard software packages offer many more functionalities than proprietary software. Moreover, the integration of various modules greatly improves the availability of data.
- The use of tried and tested standard software minimizes development time and costs, particularly compared to the development and implementation of corresponding proprietary software. On average, total IT costs of companies using standard software with few individual adjustments are around 30 percent lower than the costs of companies that primarily use proprietary software.

Total IT costs are around 30 percent lower at companies using standard software with few individual adjustments than at companies mainly using proprietary software

IT COST REDUCTION THROUGH STANDARDIZATION

MECHANICAL ENGINEERING COMPANY

A European mechanical engineering firm converted its IT to integrated commercial software between 1993 and 1995, cutting IT costs by more than 30 percent. The in-house application development and maintenance group characterized by high labor costs, obsolete COBOL knowledge, low concentration on mission-critical applications, and poor service orientation was reduced considerably. The savings in labor costs exceeded the total costs of systems implementation and operation (license fees and recurrent expenditure), calculated on the basis of at least five years' operating life by a factor of 4 (see Exhibit III–21).

Exhibit III–21
Development of Proprietary Applications versus Integrated Standard Software: A Mechanical Engineering Company

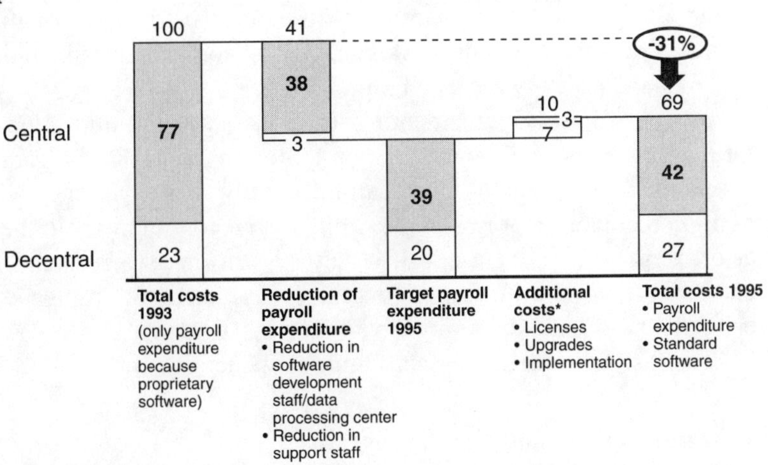

* For integrated standard software (basis: utilization period of five years).
Source: McKinsey & Company, Inc.

The potential for improvement that integrated standard software offers can be exploited only if the system is selected professionally. In systems selection, four principles should be observed:

- Profit from new software, but use a fast-follower strategy.
- Identify and exploit opportunities for systems integration.
- Choose a standard software package that optimally supports your core business processes.
- Involve IT users intensively in the selection process.

PROFIT FROM NEW SOFTWARE, BUT USE A FAST-FOLLOWER STRATEGY

Although IT stars use innovative integrated standard software much earlier than laggards do, they try to avoid being first to implement imperfect releases in their industrial sector. In 60 percent of all cases they implement standard software as fast followers—not as the first company in their industry, but much earlier than most competitors. In contrast, 15 percent of IT laggards jump at systems that have not yet passed a market test, and 70 percent decide too slowly (see Exhibit III–22).

In 60 percent of all cases, stars implement standard software as fast followers

By combining progressive IT management and smart implementation management, IT stars cut the length of their implementation projects and achieve greater effectiveness than slow followers and nonmovers among their competitors. As our research indicates, companies that consistently use the latest standard software have 20 percent better functionality and around 33 percent higher data availability than companies using older versions of standard software (see Exhibit III–23).

Instead of hastily implementing the latest software innovation as the first in their industry, IT stars bank on thorough market investigation and well-planned procurement. They carefully study user experience with initial implementation—for example, from virtual communities in the Internet or roundtables with users. If a new system offers a relevant add-on benefit for their core business processes, they decide on implementation very quickly to make the innovative functions available as soon as possible. This fast-follower strategy has multiple benefits:

- **Access to tried and tested programs**
  By waiting for the positive experience of early users, IT stars ensure that they are purchasing perfected systems that are practically bug-free and have comprehensive documentation.

Exhibit III–22
Strategies for Implementation of Integrated Standard Software by Timing of Introduction

IT stars opt for a fast-follower strategy when introducing integrated standard software*

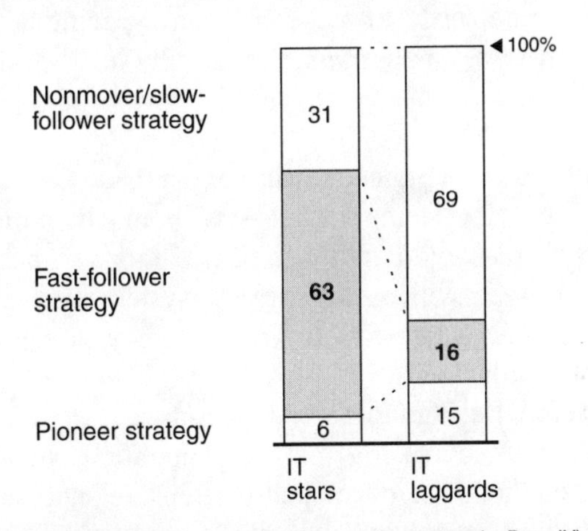

* New systems introduction (e.g., succession of proprietary system by Baan IV) and complete replacement (e.g., replacing SAP R/2 by SAP R/3), but no exchange of releases.

Source: McKinsey & Company, Inc.

Exhibit III–23
Use of Integrated Standard Software

Innovative integrated standard software leverages functionality and availability of IT

Index 0 . . . 100 (low . . . very high)

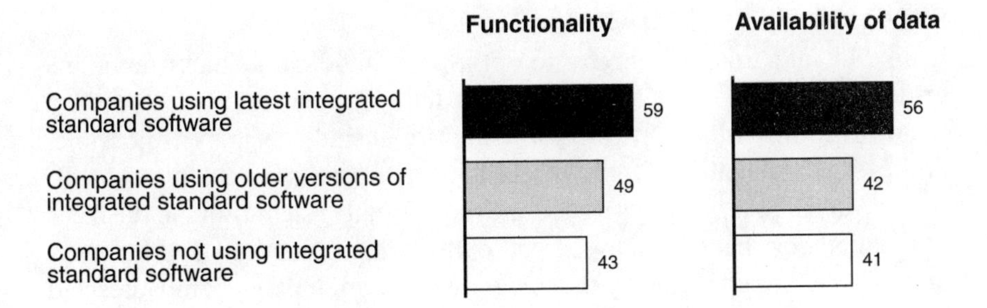

Source: McKinsey & Company, Inc.

- **Access to all systems functionalities**
  Often the first release does not yet contain all important system modules. By waiting until versions with all relevant systems functionalities are available, IT stars avoid the typical problems of first users. They do not have to develop proprietary programs for the required add-on functions or create interfaces to existing systems that already offer them. They also do not need to make company-specific adjustments to the standard software, an approach that usually leads to disappointing results and can make a later switch to more powerful releases much more difficult.
- **Availability of external consulting expertise**
  Even experienced systems analysts—without whom implementation of complex systems would hardly be possible today—need time to familiarize themselves with new software systems and develop product expertise. By choosing a fast-follower strategy, IT stars ensure that this expertise is available.
- **Use of reliable implementation methods**
  As a rule, with increasing market penetration, standardized implementation methods are developed that enable successful systems implementation in a reasonable time period. Thus, software can be implemented faster, more reliably, and in a more cost-effective manner.
  Our research reveals that IT stars especially take advantage of these circumstances. On average, the time they require for implementation of integrated standard software is 40 percent lower than that of IT laggards. Assuming the same systems operating life, IT stars enjoy longer productive use, while IT laggards lose so much time that effective use is often not possible before they have to switch systems again.

IT stars need 40 percent less time to implement integrated standard software

IDENTIFY AND EXPLOIT OPPORTUNITIES FOR SYSTEMS INTEGRATION

Unlike IT laggards, who prefer to optimize individual business functions or departments, IT stars strive to use systems implementation to create a seamless information platform for the entire business system. If possible, the new system should link all important business processes and functions. To achieve this, IT stars always determine the optimal system limit on the basis of detailed cost-benefit analyses that define which functions and business units should be supported by the integrated system. In large companies and industrial groups, the system usually does not cover more than one strategic business unit (SBU). But stars insist that the new software is the only major IT system used by all functions within the SBU (see Exhibit III–24).

Exhibit III–24
Degree of Standardization and Integration of IT Systems

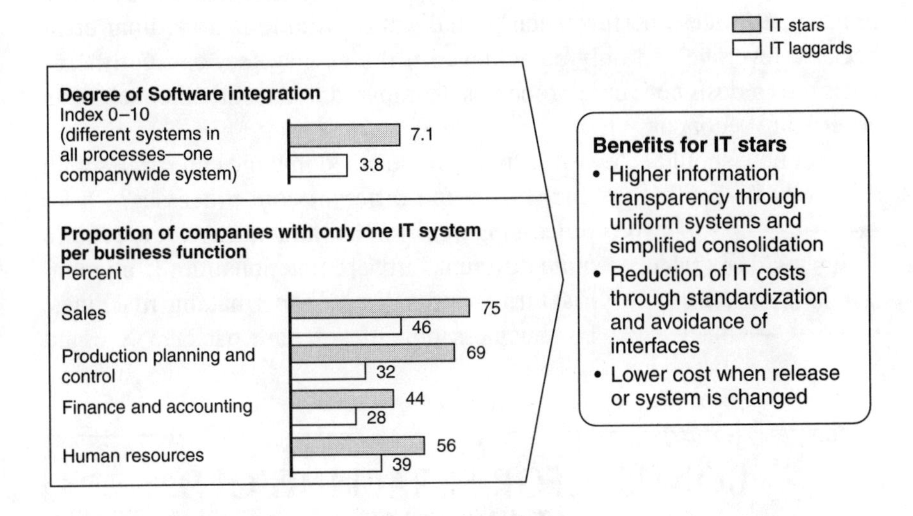

Source: McKinsey & Company, Inc.

IT laggards do not achieve such systems conformity. By using only a few modules of the integrated system and allowing the continued use of other applications, they achieve user acceptance for the integrated system more easily, but the price is high. The interface problems require substantial additional programming, and data inconsistencies frequently arise. Because the different applications do not communicate in real time, updating of the systems, at least for the programs currently used, is often possible only via nightly batch runs. Not uncommonly, important interfaces are completely missing, so that there is no seamless information flow, and data input for the different systems must be done manually.

To avoid these difficulties, IT stars prefer to make concessions on system functionalities. Our research showed that roughly two-thirds of system functionalities are used by only 10 percent of employees—and mostly only a few times a year. IT stars therefore restrict the target functionality of the integrated standard software to the 80 percent of functions that are really relevant. In special cases they provide simple Excel or Access applications as an alternative. IT stars sometimes use unconventional methods to ensure acceptance of this strategy, as the following example shows.

CHOOSE A STANDARD SOFTWARE PACKAGE THAT OPTIMALLY SUPPORTS CORE BUSINESS PROCESSES

Should a company use SAP or Baan or perhaps SSA? For many companies the name of the software vendor seems more important than the system's impact on process performance. Usually this attitude is a cardinal error because leverage can only be achieved if the new system optimally supports the (redesigned) core processes. Compared to this, all other selection criteria are secondary.

How should you select standard software?

But how should standard software be selected in large enterprises with a broadly diversified product range and different core processes? Order processing in piston manufacturing could require quite different software functionalities from manufacturing and sales of papermaking machines. For business unit A, software package X could

BONUSES FOR SCRAPPING OLD IT SYSTEMS

U.S. AUTOMOTIVE SUPPLIER

For various reasons, a U.S. automotive supplier was forced to migrate from its legacy proprietary systems to an integrated standard software package extremely fast. Resistance came from two sides. Many users blocked the switch because all user interfaces had a new design, several processes had to be changed, and not all functionalities of the old proprietary software were included in the new system. And the IT staff were against the system change because almost two-thirds of them worked in software development and maintenance and were afraid of losing their jobs if the standard software was implemented.

Management hit on a solution both simple and effective. They introduced "scrapping bonuses" for the IT staff that were paid out for each application deleted on the mainframe. The results were staggering. For the first time, there was a serious analysis of how frequently which programs were used and how the different employees in the departments benefited from them. There was great amazement on discovering how many reports were produced every month but never read. Another advantage was that IT staff were forced to familiarize themselves with the functionalities of the integrated standard software and support the new solution, because to get the bonus, they first had to win over skeptical users.

TYPICAL MISTAKES IN SELECTING SOFTWARE

U.S. STEEL MANUFACTURER

In view of the millennium problem, a U.S. steel manufacturer decided to replace its old, unadaptable proprietary software with a new integrated standard software that promised to capture substantial cost savings in maintenance and operation.

To determine which software to select, the CFO put the EDP manager in charge of a market analysis of the best-selling standard software. Three standard packages were examined closely. Because of the internal reporting structure, the project team attached particular importance to the finance and controlling modules of these systems. In the end, the CFO got a product through, against the objections of plant management, that was superior in finance and controlling but had weaknesses in supporting production planning. Initial problems already occurred in the implementation phase, because it proved impossible to integrate bottleneck-oriented production planning logic—necessary for the core business of the company—into the MRP-based system. The implementation turned out to be a complete disaster when they tried to adjust the existing planning processes in the factory to the logic of the implemented system. They could not meet delivery dates due to lack of precision and flexibility of planning; at the same time, inventory grew rapidly. Two typical mistakes were largely responsible for this development:

- The standard software was selected on the basis of its efficiency in financial controlling, even though it could not provide stable support for the company's order processing with its primary functionality. It would have been better to support order processing with the optimal software solution because this is the most important factor for corporate success.
- Management tried to adjust the rough draft of its core business processes to the requirements of the software by doing without the necessary bottleneck planning, instead giving preference to MRP II logic. This meant that in the course of software implementation, business processes were redefined in a way that did not meet business requirements. The fact that these processes were highly automated in the end did not improve the situation. One manager commented briefly and to the point: "If you are going in the wrong direction, it doesn't matter how fast you run."

look like the optimal solution, while software package Y seems ideal for business unit B. In this case, what is the best way to make a decision? In general terms the degree of standardization of IT applications will depend on two parameters:

- **Order processing in the strategic business units (SBUs)**
  Are SBUs A and B classified as standardizers, hybrid companies, or customizers (for definitions, see Exhibit II–24)?
- **Intensity of mutual supply relationships**
  Does SBU A supply parts to SBU B?

Three cases result from the possible combinations. Plausible solutions can be outlined for two of them; for the third there is no general solution.

- **Case A: The SBUs have the same order processing.**
  If all SBUs have, for example, continuous series-type production with large batch sizes and only a few customers, as automotive suppliers do, we recommend defining a single companywide standard for all SBUs, regardless of the intensity of mutual supply relationships. In this case, it is possible to develop a group standard that contains 90 to 95 percent of all functionalities. In a pilot project, a standardized implementation process can be defined and an implementation team trained that is responsible for groupwide rollout. The advantages of such a standardized approach are obvious: cost savings for procurement of hardware and software, learning curve effects during worldwide rollout, and the establishment of a compatible data basis throughout the group with supply chain management advantages.
- **Case B: The SBUs have different order processing but only limited mutual supply relationships.**
  Consider a U.S. company. One business unit manufactures simple mass products for the leisure industry, while another manufactures complex components for the space industry—on the one hand, anonymous mass production for many customers, and on the other hand, customized manufacturing for very few customers. In this case, SBUs should largely have a free hand in selecting their software. Different standard software can be used for business units with different kinds of order processing because the advantages of integration are only in joint procurement and standardized reporting to the holding company—and are therefore low in comparison to the costs of corresponding adjustments of the standard software. To enable effective financial controlling on the group level, a set of selected financial indicators could be defined that should be reported to group headquarters periodically. Technically

A CENTRAL ORDER MANAGEMENT SYSTEM INTEGRATES THE SOFTWARE OF DIVISIONS

SCANDINAVIAN CONGLOMERATE

In division A the company has make-to-order series-type production with small batch sizes, and in division B it processes raw materials in a continuous process. The raw materials partly serve as feedstock for the customized products, but are also sold directly on the market. Both divisions have the same customer group, which is served by a joint sales organization subdivided into national sales companies.

On the one hand, this company needs different production planning systems; on the other hand, the sales organization needs complete data transparency about both divisions on a day-to-day basis in order to be able to offer customers exact delivery dates. To escape this dilemma, a solution (see Exhibit III–25) was developed that consists of standard software for production planning differently configured for the requirements of the two divisions and a joint order management system for centralized rough scheduling.

In comparison to the previous stand-alone solution with different unlinked systems, the new solution offers advantages as far as data system technology is concerned—such as fewer interfaces and availability of consistent and up-to-date data—but also real process improvements. The cycle time for confirmation of orders has been reduced by 85 percent, and the time required for order processing is down by 80 percent because completeness and consistency of orders can be ensured by sales representatives who enter complete data while visiting the customer on site.

this could be done by a weekly or monthly download that is possible without the direct access of group headquarters to the IT applications of the SBUs.

- **Case C: The SBUs have different order processing and intensive mutual supply relationships.**
  In this difficult case, there is no ready-made solution. The conflict between the requirements of the group for a standardized data basis and the requirements of the SBUs for software that is fine-tuned to their own order processing cannot be solved without trade-offs. A solution developed in a Scandinavian conglomerate might be helpful.

Exhibit III–25
Concept of a Meta PPS System: Scandinavian Conglomerate

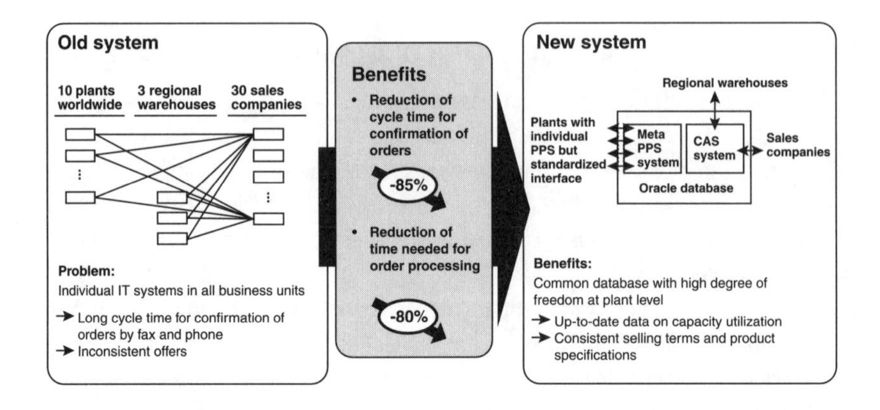

Source: McKinsey & Company, Inc.

INVOLVE IT USERS INTENSIVELY IN THE SELECTION PROCESS

To improve the utilization of IT applications, apart from periodic intensive user training, IT stars engage users in the system selection process by observing the time-honored rule for project management: Involve the stakeholders.

IT stars make sure that task assignment is clear. The IT department is primarily responsible for far-reaching standardization and integration, so it focuses on performing feasibility studies and estimating project costs. The IT users detail the targets for improvement of individual process steps that had been determined in the business process redesign phase. These detailed targets are then jointly translated into software specifications by IT staff and users.

GUIDELINES FOR PROJECT MANAGEMENT

After redesigning the core business processes and selecting the appropriate software, the fastest and smoothest possible systems implementation must be ensured. IT stars achieve this by observing the basic principles of project management:

- **Professional project planning**
  - Run pilot tests using microcosms.
  - Set ambitious targets on the basis of external benchmarks.
- **Forming project teams**
  - Assign project responsibility to IT users.
  - Focus the use of external consultants on optimizing the project concept.
- **Systematic project controlling, combined with the establishment of a process for continuous improvement.**

These principles apply to the implementation of proprietary and standard software. With standard software, a trap should be avoided that is repeatedly IT users' undoing. The basic principle for software implementation must be to avoid modifying the source code of standard software, no matter how tempting this might be! The repercussions of supposedly minor adjustments to the source code of standard software—for example, to obtain functionalities of the previous system and increase user acceptance—are generally vastly underestimated (see Exhibit III–26).

Our research reveals that companies that make considerable adjustments to their standard software have a worse cost position—their IT costs

Exhibit III–26
Cost of Standard Software at Different Levels of Software Adjustment

Modifications of the source code can be expensive

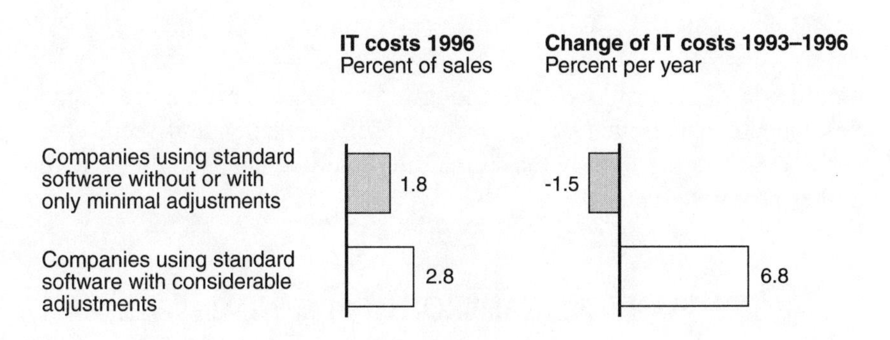

Source: McKinsey & Company, Inc.

THE REPERCUSSIONS OF UNCONTROLLED SPENDING

EUROPEAN COMPONENT MANUFACTURER

From 1993 to 1996 this company's IT costs increased by 44 percent due to uncontrolled spending on adjustments to the standard software used (see Exhibit III–27).

The manufacturer began by perfectly redesigning the business processes and then selected the most suitable standard software. However, in the detailed analysis, it turned out that many individual operations below the rough process blueprint that formed the basis for system selection did not match the software specification. Working with its software consultants, the company started to adjust the source code of the standard software in order to tune the software to the work flow. The result was an extremely long and expensive implementation project. Most serious were the long-term problems that resulted from the high degree of software adjustment. Every time that upgrades of the standard software were released or accompanying applications changed, the sequences that had been modified in the source code had to be adjusted. This was so complicated that experts from the software vendor had to be called in, causing costs to soar.

It would have been much better for the company to stick with defining only the rough blueprint of the business processes during the redesign phase and then adjust the fine detail of the individual operations to the program requirements after software selection.

are at least 50 percent higher. The long-term impact is even more disconcerting. Companies that adjust standard software may find themselves trapped in spiraling cost increases that ultimately destroy their leeway for further IT investments.

PROFESSIONAL PROJECT PLANNING

IT stars pay great attention to careful, comprehensive planning of the migration from the old to the new IT system. Two-thirds of their total manpower invested in implementation is used for planning, whereas this

Exhibit III–27

Cost Increase Due to Software Adjustment: Component Manufacturer

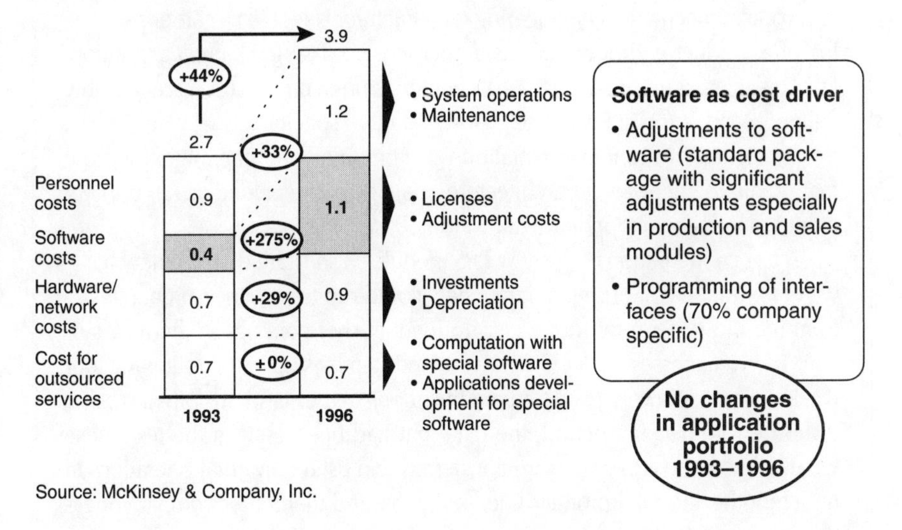

Source: McKinsey & Company, Inc.

Two-thirds of stars' total manpower for implementation is used for project planning, versus 50 percent at laggards

figure for IT laggards is only around 50 percent. In the planning process, IT stars focus on the concept for the project approach, specification of objectives (project duration, milestones, costs, and responsibilities), and deciding on external consultant involvement in implementation.

RUN PILOT TESTS USING MICROCOSMS

IT stars prefer a step-by-step implementation. Integral to this approach are pilot tests of individual modules of the new software in restricted user environments. Usually such microcosms are a complete core process. Only after the pilot tests have been successful will there be a rollout for major business functions or units. Using this approach, IT stars can make best use of the advantages of the sequential implementation strategy, as well as rapid prototyping of software and modular implementation.

PILOT TEST IN A MICROCOSM

MECHANICAL ENGINEERING COMPANY

A European mechanical engineering manufacturer selected the order processing of a product with low sales as a microcosm for systems implementation. For this clearly defined test area, new business processes were roughly defined, and after software selection detailed specifications were made. Because the software implementation was combined with the migration from mainframe to client/server architecture, management decided to run a three-month parallel operation in the transition phase.

A test configuration of the new PPS module was installed on the new hardware. On that basis, production planners could run tests with real data (taken from the old system) for four weeks to identify shortcomings in the new system. The weaknesses detected were resolved step by step in the following four weeks using a few very restricted modifications of the standard software. After a short error checking period, the data that had been used in the test phase were deleted; the real data were transferred and used in normal operation. In nine months, the total product line, the plant, and then other company locations were gradually switched over to the new system.

Analyzing this approach from the viewpoint of project management, we see that practically all implementation tools were used: sequential strategy, rapid prototyping (running a test on the basis of a minor but typical product), and modularization (starting with PPS, and then rollout to other modules). Thus, the benefits of the different methods could be used, including swift progress, limited project risk, and intensive involvement of IT users.

SET AMBITIOUS TARGETS ON THE BASIS OF EXTERNAL BENCHMARKS

Based on the overall business objectives pursued in systems implementation, IT stars set detailed targets for system functionalities, project duration, project costs, responsibilities, milestones, and indicators for review of operating, maintenance, training, and other costs incurred as a result of implementation. After all, "If you don't take any measurements, you can't expect any improvements," commented a French top manager, briefly and to the point.

It does not make much difference which project planning system this target-setting process is supported by; we could not identify a clear winner. The factors that really seem to make the difference are rigorous execution, checking of target fulfillment in reviews, and the aspiration level set by management. For IT stars, agreements on targets are always definite promises that are binding for management and staff, the IT department, and users (see Exhibit III–28).

The overriding objective is to develop internal customer-supplier relationships that offer IT users clear horizons for their planning and at the same time allow management to evaluate IT department performance as objectively as possible. Following these guidelines, IT stars do not simply estimate technical and commercial objectives internally, as most IT laggards do (e.g., system functionalities versus implementation and recurrent costs), but often use external benchmarks to achieve high-quality solutions at minimum implementation costs and project duration. The targets, derived from benchmarks and defined together with employees, are mostly put in writing to create clear ground rules for everyone involved.

FORMING PROJECT TEAMS

Usually implementation teams are successful only when IT users are involved early and clear rules for the involvement of external consultants are defined. Experience has shown time and again that overall responsibility

Exhibit III–28
Setting Targets for IT Projects

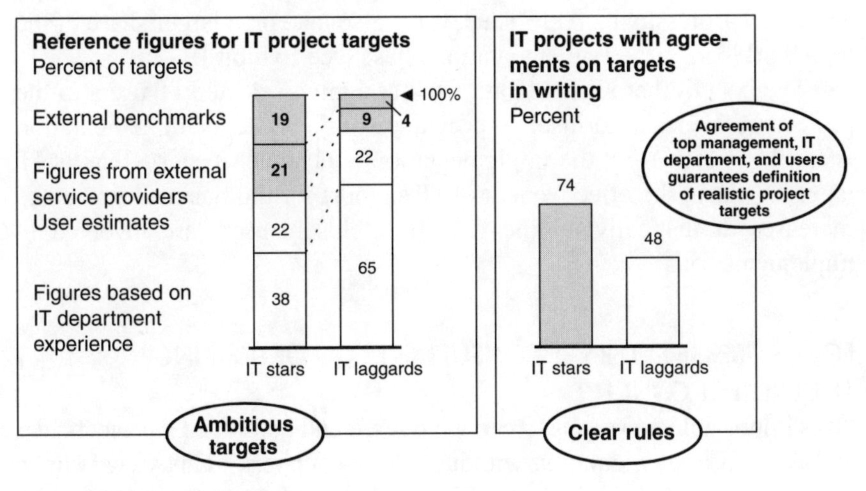

Source: McKinsey & Company, Inc.

COMPETITIVE BIDDING AS BEST PRACTICE

SCANDINAVIAN CONGLOMERATE

A Scandinavian conglomerate put out a public tender for the extension of its intranet to create a competitive situation between the in-house IT department and external service providers. The top management executive committee decided that the project should be performed internally only if their own IT department's quote was at least comparable to the best external bid. This way they smartly circumvented the common practice of IT departments of estimating targets based only on experience, with a generous productivity buffer to ensure achievement of the planned targets.

for the implementation project is best assigned directly to IT users, confining the use of external implementation consultants to concept development.

ASSIGN PROJECT RESPONSIBILITY TO IT USERS

IT stars not only involve users in the software selection process, but also assign them project management in the implementation phase. Preferably these are the lead users, discussed in Rule 6. If problems arise due to insufficient know-how, a tandem arrangement is recommended. Instead of one project manager, a management duo is appointed consisting of an IT user and an IT professional. At IT stars, users also take on a larger share of the total workload in the implementation phase (see Exhibit III–29).

The benefit that IT stars gain from assigning selected IT users to the project work speaks for itself. Acceptance of the systems implementation is much greater—and the implementation itself much faster—than at IT laggards. Also, the effectiveness of IT support for the business processes increases dramatically in line with how closely users are involved in implementation.

FOCUS USE OF EXTERNAL CONSULTANTS ON OPTIMIZING THE PROJECT CONCEPT

"It is impossible nowadays to carry out large and complex IT projects like an SAP R/3 implementation without external support." This view is held, our interviews showed, by both IT laggards and IT stars. But assessments

Exhibit III–29
Benefits of Intensive User Involvement in IT Projects

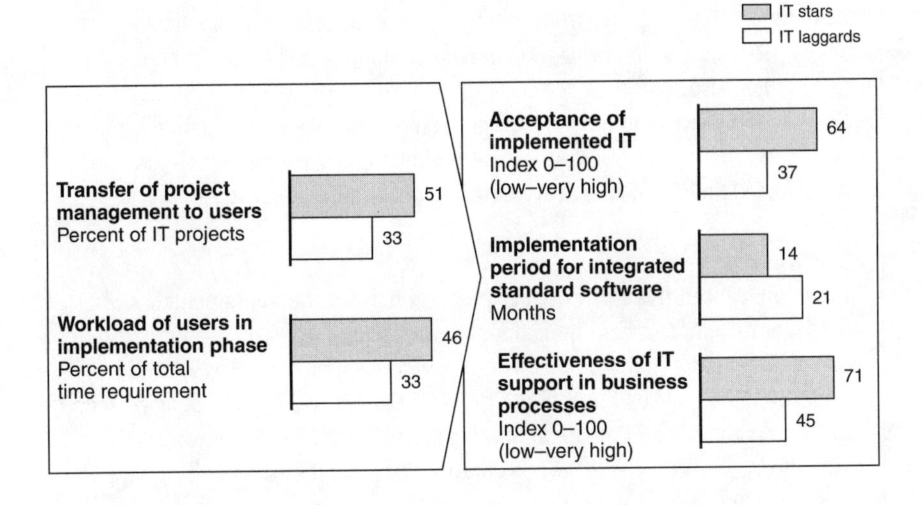

Source: McKinsey & Company, Inc.

of the performance of external consultants diverge widely. IT laggards especially complained about difficulties in cooperation. Criticism ranged from “rip-off mentality” and “unable to understand our processes” to “unqualified.” We rarely heard statements like these from IT stars. How do IT stars succeed in making better use of external consulting expertise? What is it that they do differently?

We found that IT stars tend to use external consultants for implementation more intensively than laggards do—but with completely different expectations. For stars, the goal is to acquire the expertise of the consultants as quickly as possible and then to become independent of them again just as quickly (see Exhibit III–31). The examples in the case study regarding the use of external consultants also show how vital it is to exempt project team members as fully as possible from their day-to-day activities and to integrate them properly into the project.

Even in the concept phase, IT stars call in external consultants to check the proposals they have developed in-house or to get information about the latest IT products available on the market. For implementation support, IT stars extensively use the consulting services of software vendors, while IT laggards are more likely to cooperate with independent IT consultants.

USE OF EXTERNAL CONSULTANTS

At company A (an IT laggard) the implementation of a well-known standard software took six years, twice the time of company B (an IT star), although A is far smaller. Also, company A's costs for external consulting and implementation support were a multiple of the figure company B spent on external service providers. These deviations are due to the different degree to which the companies used their own staff in systems implementation (see Exhibit III–30).

- **Company A**

Company A used five of its own IT professionals for the implementation, but they were exempted from their normal work only 50 percent and could not play a driving role in the project. The staff were supported by 40 external consultants. In the course of user training, 70 IT users were assigned to the project for 10 percent of their time to familiarize themselves with the software system. This allowed them to make limited contributions, but they did not have time to scrutinize the solutions closely and evaluate the different options—to say nothing of perhaps reconsidering the work flows from a company perspective.

- **Company B**

Company B exempted five IT professionals and eight IT users fully from their normal work and used only two external consultants for the pilot project. The objective of company B was not only to introduce the system in the pilot area, but also to familiarize employees with the implementation management. The pilot project was deliberately restricted to 12 months. The global rollout was conducted in 12-month intervals by internal staff.

IT stars always pay attention to defining the assignment clearly and make sure the consulting services can be precisely controlled. If they have developed sufficient expertise of their own, they are anxious to restrict the participation of third parties in implementation to a minimum. This prevents possible conflicts of interest and avoids the negative experience that seems all too familiar to laggards: "Our consultants in the PPS project are interested only in selling man-days, not in pushing implementation" was the way one IT manager at a laggard summed up his experience with external consultants during implementation.

Exhibit III–30
Use of External Consultants for Systems Implementation: Companies A and B

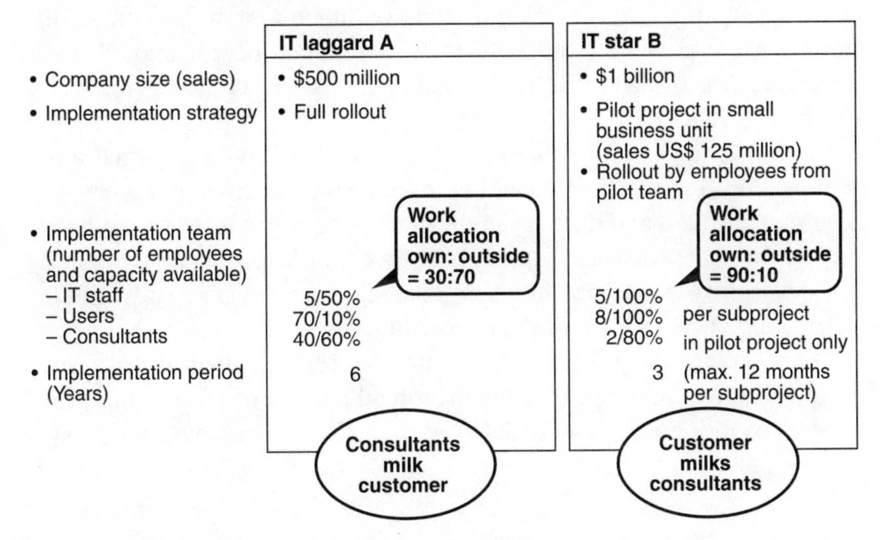

Source: McKinsey & Company, Inc.

Exhibit III–31
Manpower Structure for Planning IT Systems Implementation

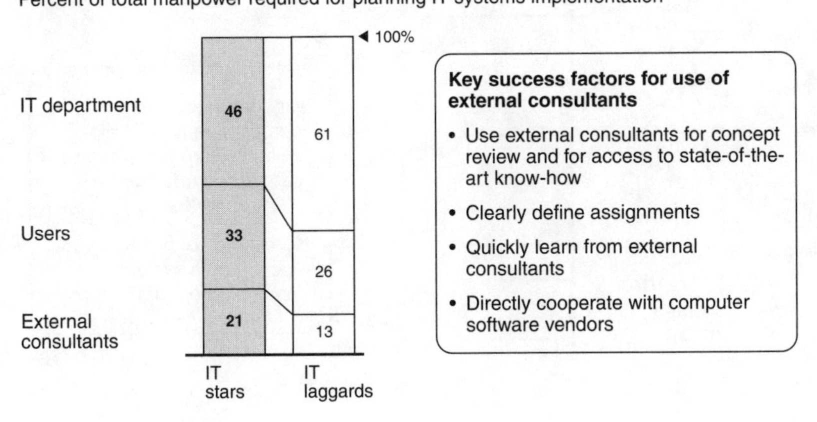

Source: McKinsey & Company, Inc.

SYSTEMATIC PROJECT CONTROLLING

What is taken for granted in other functional areas and product development projects has not yet become a standard in project management at many IT laggards and also at some IT stars: resolute project controlling combined with follow-up reviews and continuous improvement. As the survey results indicate, the problem is insufficient organizational discipline rather than a lack of IT tools for supporting project management (see Exhibit III–32).

It is true that more than 80 percent of IT laggards use formal milestone controlling of IT projects, but top management merely reviews the accumulated costs and time schedule, playing no active role in steering the project on content issues. This indifference can largely be explained by the low priority that IT has among top management at IT laggards, and the insufficient IT know-how of many board members.

Our analysis of project review activities revealed that even IT stars need to improve in this area. Only the top 30 percent of IT stars use semiannual postmortem reviews to determine whether the implemented IT sys-

Exhibit III–32
Parameters Used for Project Reviews

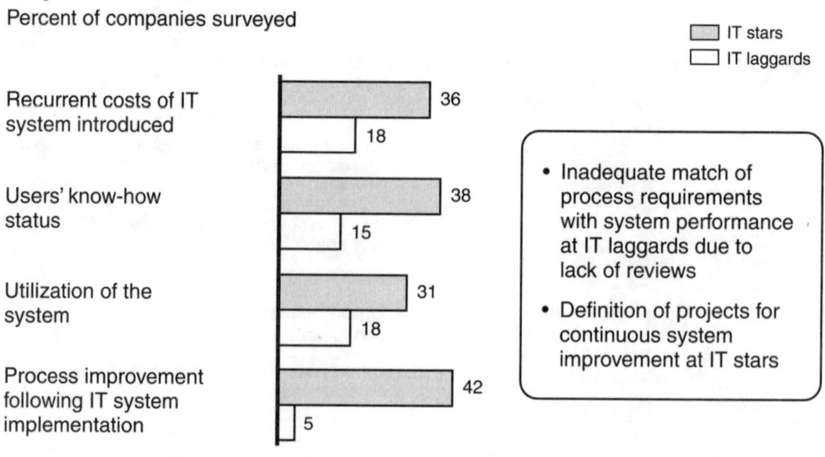

Source: McKinsey & Company, Inc.

tem is running within its budget, whether IT users have been trained as planned, and whether utilization corresponds to original expectations. The others—and especially the IT laggards—risk a rude awakening. In subsequent audits, it is often discovered that operating costs are unexpectedly high, and the software implemented is not nearly as user friendly as anticipated. The hot line is being used more than planned, and user acceptance is so low that the old application is enjoying a shadow existence alongside the new one.

IT stars develop a continuous improvement process for their IT implementation management

IT stars strive to take countermeasures and check regularly whether and to what extent improvement is needed (see Exhibit III–33). They record their experience in project databases and similar files to learn from mistakes and use positive input for subsequent projects. This enables them to develop a continuous improvement process for IT, similar to the systems that have sprung up for production since the introduction of quality management.

Exhibit III–33

Management of IT Systems Implementation: Virtuous versus Vicious Cycle

IT stars continuously improve their IT implementation management

IT stars: Upward spiral

IT laggards: Downward spiral

		IT stars	IT laggards
I	Top-management aspirations	High/ambitious goals	Low/less ambitious goals
II	Degree of goal achievement (deviation)	High (13%)	Low (42%)
III	Implementation period for integrated standard software	14 months	21 months
IV	Satisfaction with process (index 0–5)	4.1	2.9
V	Aspirations for later projects	Even higher	Even lower

Source: McKinsey & Company, Inc.

SUMMARY

Mastering technological progress is still the greatest challenge for implementation management. But major implementation projects do not have to become a nightmare. The key is to follow the sequential strategy of IT stars—first business process redesign, then systems selection, and finally implementation—and observe the success factors of the different steps (see Exhibit III–34).

The limiting character of technology in itself is likely to decline, driven by four key trends identified in our research:

- **Open systems architecture of standard software**
  As a consequence of this trend, it will be easier to combine different modules, and so IT penetration of core business functions will accelerate. This will offer new opportunities to achieve competitive advantage.
- **Microprocess modeling techniques**
  Techniques such as business engineering workbench and dynamic enterprise modeling will shorten implementation processes considerably and change the demands on implementation teams.
- **Microprocess libraries**
  With the development of industry-specific and business-process-specific microprocess libraries (also business repositories and enter-

Exhibit III–34
Implementation of Complex IT Applications: Summary

Key principles for successful IT implementation

	Business process redesign	System selection	System implementation
	Redesign business processes before systems implementation; innovative software does not guarantee efficiency and effectiveness	Introduce innovative application systems early	Observe principles of project management and perform pilot tests in microcosms
Success factors	• Confine yourself to specifying only blueprints for business processes • Involve internal IT staff and external strategy consultants early • Set clear improvement targets based on external benchmarks • Use innovative IT functionality for optimizing business processes	• Use pioneer strategy only when IT can give you a competitive edge • Introduce integrated standard software as fast follower – Benefit from new software, but wait for market acceptance – Determine and fully exploit the potential for integration – Focus on core business processes – Involve users intensively in the selection process	• Principle: Don't modify the source code in implementing standard software • Use professional project management – Perform pilot tests in microcosms – Set ambitious goals on the basis of external benchmarks • Form project teams – Users should be in charge of project management – Use external consultants to optimize the concept • Establish systematic project controlling

Source: McKinsey & Company, Inc.

prise reference models), business operations will increasingly become standardized. In the medium term, it will be possible to implement best-practice processes faster and more easily.

- **Project management as standard know-how**
  Cross-functional companywide teams and the alignment of organizational structures to core business processes will make project management standard know-how for most employees.

The relative importance of the different steps in the IT implementation process will change as a result. Systems selection and implementation will become less critical, while business process redesign will gain even greater significance. This will mean IT users must be involved even earlier and more intensively in this phase of IT projects.

At the same time, professional IT management is growing in importance. It must be able to evaluate and communicate the opportunities of new IT developments, while initiating and following through on the technological and organizational adjustments required.

Part IV

OUTLOOK ON THE FUTURE OF IT

Ultimately a survey can only ever be a snapshot of the status quo. Certainly we can use it to derive recommendations for the future. But how can we be sure that these recommendations will maintain their validity in the light of the extraordinary dynamics of IT? New technologies and applications could conceivably overturn all previous ground rules, creating completely new paradigms for best practice. Or could the trend toward standardization so vastly simplify IT management that users take decisions into their own hands? If technology becomes sufficiently streamlined, you can imagine the sales department and its users, for example, making their own choice on which hardware and software to buy.

These possibilities do not seem realistic. We are convinced that at least for the foreseeable future, the interaction between growing standardization and the dynamics of the IT market will lead to further complexity, thus increasing the demands on IT management. In view of this, the recommendations of this book on superior IT management will become all the more important. In the long term, we see the gap between IT stars and laggards becoming even greater if countermeasures are not taken. The stars will find they can build on their lead exponentially, while laggards lack the requirements needed to catch up.

To conclude this book, we examine the trends and interrelationships that we believe will shape the future in this field. Starting with an overview of IT dynamics, we go on to illustrate future developments relevant to the manufacturing sector and close with their likely implications for IT management.

DYNAMICS OF IT DEVELOPMENT

Due to its rapid speed of change, no more than a rough outline can be drawn of the future development of IT and the consequences for IT management. Moreover, to date virtually no research has been conducted on the interaction of the multiple factors influencing this development (see Exhibit IV–1).

From today's perspective, three primary factors are likely to drive the complexity of IT management:

- Emergence of new base technologies, enabling completely new IT applications in a cyclical pattern
- Interaction between technological change and standardization
- Substitution of existing applications and technologies due to complex market requirements

These factors will dramatically offset the potential for simplification that results from standardization and accumulated experience.

Exhibit IV–1
New Challenges to IT Management

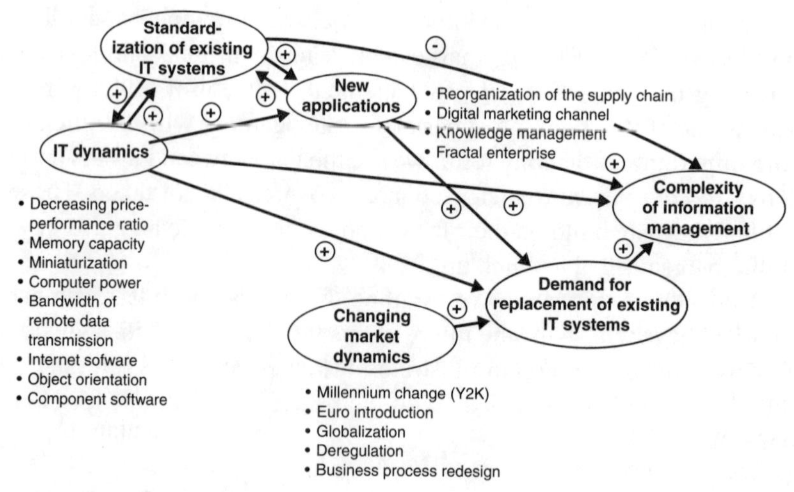

Source: McKinsey & Company, Inc.

EMERGENCE OF NEW BASE TECHNOLOGIES

The development of IT is characterized by a continuing trend toward higher performance, combined with a dramatic price decline. In the last seven years alone, the cost of one hour of computing center time has fallen by 90 percent. The switch from mainframes to open networks and PCs is a further driver of this trend. The performance of Intel Pentium chips rose by a factor of five from 1995 to 1998, and is expected to double again by the year 2000. Communication costs will also continue to fall. This development will be accelerated by the ongoing deregulation of telecommunication markets in Europe, among other factors. Transmission bandwidth will increase from around 64 kilobytes per second or 34 megabytes per second (in special networks) today to more than 100 megabytes per second in the long term. This combined with plummeting costs will make the Internet the medium of choice even for data-intensive forms of transmission such as videoconferencing.

Internet software technologies such as HTML and Java make it possible for companies to develop and operate new application systems at low cost and high flexibility in-house. In the near future it will be possible for secretaries to develop their own intranet applications by designing HTML pages (e.g., for travel reservations). Internet service providers such as Netscape already offer complete programs on this basis for handling travel expenses or for inventory management. According to IDC, the implementation costs per user amount to $1,500—with a payback period of sometimes only six to twelve weeks.

Vendors of special-purpose software, such as Actra from Silicon Valley, offer programs to bring together customers and suppliers in the World Wide Web. They integrate functionalities from database systems, search engines, security software, and payment systems into a toolbox, making it possible to adjust the transaction process to customer needs.

In the United States, 63 percent of major enterprises already use or plan to use intranets. According to estimates by the market research institute Durlacher Multimedia, by 2001 20 percent of IT expenditures in the United Kingdom is likely to be for intranet applications.

Since the end of the 1980s, computer-aided software engineering (CASE) tools have allowed systematic documentation of software development. Nevertheless, their impact on the productivity and quality of software has remained limited because adapting models developed with CASE was often very time-consuming, and direct implementation of models was not possible anyway. But for several years, so-called object-oriented models for modeling microprocesses (OPM) have been available that directly describe microprocesses in object-oriented programming languages (e.g., C++). These overcome the previous disadvantages of CASE. All major vendors of integrated standard software now use OPM technology (e.g., Baan) or are planning to introduce it. This can already reduce the costs of introducing standard software to two to three times the license costs, rather than the previous factor of 5 to 10. In the medium term, the automatic installation of complex integrated application systems could even become a reality.

Object-oriented programming languages could also challenge the integration benefits of integrated standard software. These languages make it possible to describe complex interfaces between different areas of application on the level of so-called business objects. This can greatly simplify the communication between the individual components of integrated standard software and other IT applications that was always so complex.

Software vendors are working to develop the next generation of programs based on object-oriented systems architecture. The goal is to define a new interface standard on the level of business objects. As soon as companies begin to adopt this new systems architecture for integrated standard software, we can expect a new wave of supplements and refinements to the programming of business components ("component software"). The reason is that as individual components are modified and improved, interface costs no longer play such a significant role, thus stimulating the incentive to capture individual optimization potential.

INTERACTION BETWEEN TECHNOLOGICAL CHANGE AND STANDARDIZATION

Parallel to the development of new base technologies and IT applications, the same forces will also drive the standardization of existing base technologies. This will expand the accessible market as prices decline and performance increases—reason enough for software developers to want to exploit the cost advantages inherent in product standardization or the development of an industry standard. This is how the UNIX standard for operating systems of midrange computers developed in the 1980s. Moreover, so-called network economies favor the emergence of standards or quasi-standards in information technology. For example, the benefits for users of an Internet service provider or Internet community site increase exponentially with the number of users.

Network economies fuel the emergence of standards or quasi-standards

But standardization will not result in a commoditization of information technology. The evolution of IT can be compared to that of road traffic. Building up a standardized road infrastructure creates the incentive to develop more vehicles. The new range of vehicles does not simply lead to a shift in existing traffic but creates additional traffic volume, which in turn stimulates further infrastructure expansion. In this sense the standardization of existing IT systems stimulates expansion of the spectrum of IT applications through economies of scale, and greater demand for IT. This generates new base technologies and applications based on emerging standards. With Windows and Intel microprocessors, Microsoft and Intel created quasi-standards for PCs that have motivated many small- and mid-sized software firms to develop special-purpose application software for new user groups. Similarly, with the rise of SAP, a large market has sprung up for SAP implementation support and consulting services.

SUBSTITUTION OF EXISTING IT APPLICATIONS AND TECHNOLOGIES

New, powerful IT applications and technologies create a constant incentive to accelerate the replacement of existing platforms and applications. Industry trends such as globalization or deregulation—if they stimulate adjustments in corporate strategy and related improvements to business processes—can make existing systems obsolete and accelerate their adjustment or replacement. The same applies to special events such as the introduction of the euro, or the millennium bug that has triggered a tidal wave of software replacement.

The dynamics of technology and markets drive the premature substitution of IT systems, making the migration of existing systems a central challenge for IT management. Even if the degree of standardization grows, at the same time entirely new technologies and areas of application are emerging that hugely expand the spectrum of IT applications in the manufacturing sector. This in turn fuels the complexity of IT management.

The dynamics of technology and markets drive premature substitution of IT systems

NEW IT APPLICATIONS

The extent to which the areas of application of information and communication technologies will change can be illustrated along two dimensions (See Exhibit IV–2).

Exhibit IV–2
Key Applications of Information and Communication Technologies

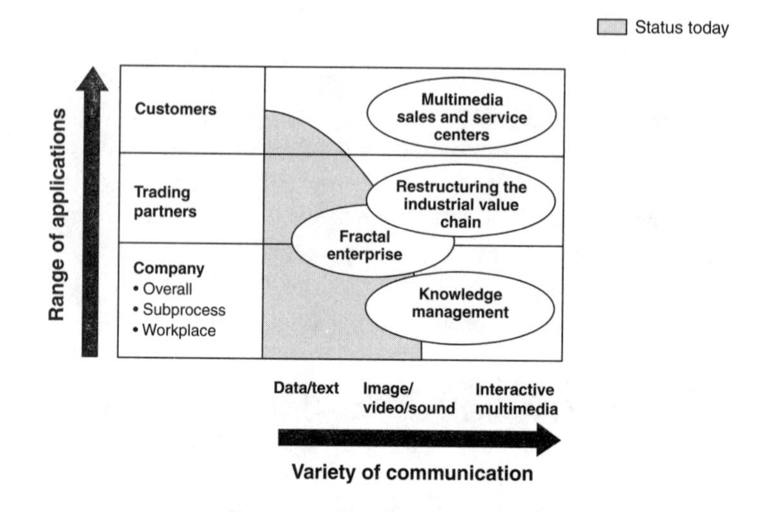

Source: McKinsey & Company, Inc.

In the near future the variety of information and communication will soar as a result of IT. Primarily a medium for data and word processing in the past, today IT is making the processing of audio and visual data ever simpler and more interactive (e.g., PC video conferences). At the same time, the range of applications will increase. Up to now, the processing of images and sound required expensive special-purpose computer networks, but with Internet technology, a cost-effective platform is now available—for communication both in-house (intranet) and with partners and customers.

We will now examine four key potential areas of application for information and communication technology.

INTEGRATED KNOWLEDGE MANAGEMENT

"Knowledge is information combined with experience, context, interpretation, and reflection—that is to say a high-quality form of information that is always available for decisions and actions." Based on this well-known definition by Thomas H. Davenport, a discussion has developed in recent years on how the individual and organizational *knowledge* that

THREE-LEVEL MODEL OF KNOWLEDGE MANAGEMENT

Three levels of IT-supported knowledge management can be distinguished.

- **Level 1**

On the basic level of knowledge management, the primary intention is to make explicitly available, structured knowledge available. This classification applies to all projects in which different reports (e.g., R&D reports, call reports of sales representatives) are collected electronically, or market and competitive information from various sources (e.g., the Internet) is gathered and sifted.

When building these knowledge databases, it is essential to use suitable classification systems, filter processes, and efficient search algorithms. Of key importance is the combination of conventional data (e.g., technical product information) with case-specific documents (e.g., marketing-oriented product information), as well as the use of intelligent filters, with which information can be allocated to specific subjects in a situative or event-oriented manner.

- **Level 2**

On the next level of knowledge management, IT-supported access to explicit, structured knowledge can leverage further significant potential. Groupware tools such as Lotus Notes are used in combination with global telecommunication networks and specific options for document management. Typical applications are: reusing parts in product design, sharing experience about research results, and creating "who's who" directories for experts in specific areas. For example, using a knowledge management system of the second level, Microsoft developed an internal electronic job market to find the best-qualified employees for special projects.

- **Level 3**

The most ambitious application of knowledge management is to transform unstructured knowledge only available to individuals into explicit, structured knowledge that is available to the whole organization. The task is not simply to digitalize implicit knowledge. A comprehensive organizational approach is necessary that outlines the structuring logic for the processing and evaluation of these data, allowing access to what was previously exclusive knowledge. If this approach can be institutionalized, it will greatly encourage the development, dissemination, and leverage of knowledge. Entirely new ways of organizing and assigning work in companies will emerge, and the cooperative use of knowledge will gradually substitute exclusive, unshared knowledge.

exists in institutions can be systematically tapped and refined using state-of-the-art information and communication technology.

This discussion was especially stimulated by the development of specific data processing standards for text, images, and sound, as well as the spread of groupware and intranet tools that are nowadays available via the Internet. Meanwhile, progressive IT stars have started to use these IT applications as platforms for gathering and disseminating knowledge in corporate networks in the new drive for "knowledge management."

Obviously on the path to cooperative use of knowledge, IT can only provide an appropriate transfer platform. To successfully implement the higher levels of knowledge management, a more holistic perspective is needed that is also reflected in the company's human resources management and organization structure.

All these aspects make effective knowledge management a complex corporate task. The essential *knowledge goals* of the company must be anchored as guiding concepts within the scope of strategic planning. To achieve these goals, they need to be broken down into concrete operational objectives, with defined organizational ground rules for gathering and disseminating knowledge. The IT-supported knowledge management processes supplement or compete with various informal staff networks. Workshops, meetings in the cafeteria, and telephone calls are still the most important media for informal knowledge transfer. Perhaps the increasingly popular intranet discussion forums for selected user groups will be the vehicle that helps IT-supported knowledge management to make a breakthrough.

A company's key knowledge goals must be anchored in its strategic planning

However, selfish motives such as "knowledge is power" or the familiar "not invented here" syndrome often block the dissemination and use of knowledge. Without an open knowledge culture that generally stimulates the evolution and exchange of knowledge in organizations and promotes mutual learning, it will not be possible to harness the full potential of the new information and communication technologies.

When knowledge management is understood as a change program, it can have a major impact on companies. Stars welcome the challenge of integrating inherent knowledge from different areas of the company via IT.

A system developed by Motorola (see accompanying box) illustrates the possibilities of integrated knowledge management.

KNOWLEDGE MANAGEMENT APPLICATION AS PART OF A WORK FLOW MANAGEMENT SYSTEM

MOTOROLA

For the manufacture of modems with complex assembly and sophisticated quality standards, Motorola installed a knowledge management application on the shop floor level that is a pilot for other product lines. In product development, the individual components of the modem are first digitalized by a digital camera. Then the pictures are enhanced with detailed instructions for assembly, testing, packaging, and shipment. At the assembly line, employees see the assembly instructions on screen. Special marks help them to identify new parts and the corresponding instructions. They can feed details of faults and their causes into the computer directly at the assembly line. Via intranet, the documentation is modified on the web server and is simultaneously made available to all other job terminals.

The Motorola intranet accelerates the improvement process when products are launched or adapted. It is also an excellent way to meet ISO specifications more efficiently. The costs for developing this system were relatively low, but it is important to have strict rules for using the system to make lasting improvements to the quality and consistency of the manufacturing process.

THE FRACTAL ENTERPRISE

The systematic use of conventional IT—internally, but also in communication between the company and its partners—is an opportunity for fundamental company reorganization. Buzzwords such as *functional outsourcing, fractal, modular* or *virtual enterprise* describe different aspects of a trend that will become ever more important in the years to come. IT systematically promotes competition at different stages of the value chain (or supply chain), stimulating the development of partner networks. Companies that take advantage of this trend can sustainably improve their competitive position.

It is easiest to describe how IT influences organization structures and competition along the value chain by looking at key trigger mechanisms. We will examine three of these as examples (see Exhibit IV–3).

Exhibit IV–3
Impact of IT on Company Organization

IT stimulates competition on the different levels of the value chain

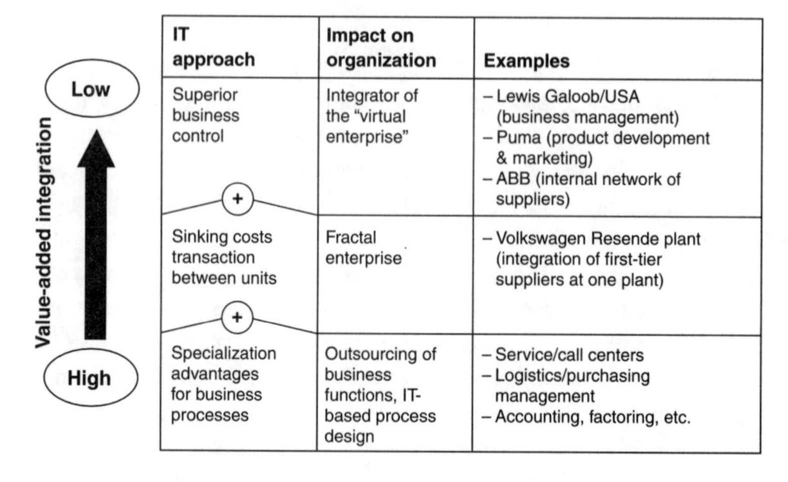

IT approach	Impact on organization	Examples
Superior business control	Integrator of the "virtual enterprise"	– Lewis Galoob/USA (business management) – Puma (product development & marketing) – ABB (internal network of suppliers)
+		
Sinking costs transaction between units	Fractal enterprise	– Volkswagen Resende plant (integration of first-tier suppliers at one plant)
+		
Specialization advantages for business processes	Outsourcing of business functions, IT-based process design	– Service/call centers – Logistics/purchasing management – Accounting, factoring, etc.

Source: McKinsey & Company, Inc.

IT CALL CENTERS

IT call centers allow real performance leaps in serving customers by telephone. IT-supported automatic call distribution (ACD) minimizes the waiting time for customers. Simple inquiries can be handled by IT expert systems (touchtone systems/fax-back services). The information and advice call centers will continue to improve—even extending to telediagnosis and on-line support to service and sales field staff. This will be possible due to sophisticated consulting software and availability of highly accurate customer information.

IT-SUPPORTED REORGANIZATION CAPTURES SPECIALIZATION ADVANTAGES

Utilizing specialization advantages on different steps of the value chain via IT-supported reorganization is not new. As early as the 1960s, bookkeeping and payroll accounting were fundamentally reorganized by the introduction of mainframe computers. Specialist service providers successfully established themselves in this market. In the past ten years, the areas suitable for IT-based reorganization have expanded to cover all functions, from specific development tasks (e.g., rapid prototyping), through logistics services (e.g., inventory management) and sales and service (e.g., invoicing). The most recent example is the reorganization of sales and service tasks in call centers.

Usually only a superior concept for a new process, not purely the technology, leads to successful reorganization. But IT plays a key role; understanding its potential is vital for developing and then realizing the concept.

In spite of the many special providers who offer to take over operational functions for almost all steps of the value chain, careful analysis is needed to decide whether it is sensible to outsource processes or to reorganize them internally. Our research shows that IT stars recognize the potential for outsourcing early, and then make well-considered decisions. They use external service providers primarily when the process or function is not a strategic core competence and outsourcing will significantly raise the performance of the process in question.

IT-BASED NETWORKS REDUCE TRANSACTION COSTS AND PROMOTE SYSTEMS PARTNERSHIPS

We have already discussed the benefits of using IT to integrate customers and suppliers. Solutions that are based on common technologies such as CAD data exchange and EDI-based purchase order processing can substantially reduce transaction costs between business partners. They can also fundamentally change the relationship between manufacturer and supplier. For the manufacturer, it may prove beneficial to outsource additional steps in the value chain to suppliers (such as product development or assembly activities). Alternatively, the manufacturer may decide to take on tasks previously performed by customers. Taking this to the extreme, a fractal factory emerges in which different systems partners work closely together at one manufacturing location, as the following example illustrates.

Realizing a model of fractal manufacturing requires more than strategic analysis of the business system. IT management is also confronted with new challenges. The new IT infrastructure must be aligned exactly to the new business processes, from procurement via production and operations management (POM) to shipment. A superior IT concept is needed as well as professional project management to overcome the considerable risks in the transition phase.

THE FRACTAL FACTORY

VOLKSWAGEN (RESENDE PLANT IN BRAZIL)

The Volkswagen plant in Resende/Brazil illustrates a completely new type of organizational model.

In 1996, Volkswagen built an assembly plant for trucks and buses with 1,500 employees, of whom only 200 work directly for Volkswagen. Eight completely independent systems integrators assemble their components (e.g., gearboxes, axles) on the same site. The systems integrators closely cooperate with Volkswagen to align the different components in the predevelopment phase. Each systems partner delivers "just in time" according to an agreed logistics concept.

The tasks start with operations planning. The suppliers must observe high-quality standards and meet delivery deadlines. Controlling this complex assembly process requires not only high reliability and quality standards on the part of the systems partners, but also a highly sophisticated IT infrastructure. Without adequate IT support, such a highly integrated logistics and manufacturing process would be impossible.

IT-BASED COORDINATION ENABLES SUPERIOR BUSINESS CONTROL

The trend toward declining transaction costs facilitates the outsourcing of individual value-added steps (e.g., product development, production or sales activities). The company can focus on developing and marketing its business idea and coordinating the value-added process. Taken to an extreme, work with individual partners takes place on a project-by-project basis, according to the partners' particular strengths (as a "virtual enterprise"), without the complicated and sometimes exclusive framework agreements of systems partners.

There are already a few examples of such virtual enterprises. For instance, Lewis Galoob Toys, a U.S. manufacturer of toys, focuses on managing a network of independent companies and marketing the goods. This company acts as a systems integrator. It buys product ideas and product development services from independent inventors and engineering offices. Subcontractors in Hong Kong and China are in charge of manufacturing. Logistics and sales are performed by independent forwarding and sales agencies.

Falling transaction costs will increasingly lead to outsourcing of steps in the value-added chain

Puma, the German sports equipment manufacturer, had no choice but to develop into the integrator of a partner network by rigorous outsourcing and focus on core competencies. The company found it had serious performance deficits in the production of several product lines. It retained only product development and marketing as core business functions, and is also considering outsourcing logistics entirely.

Virtual enterprises need an efficient information and communication infrastructure to ensure that decentralized tasks and distributed knowledge are coordinated. The use of IT for superior business development and control thus becomes a crucial competitive advantage.

Competing with conventional manufacturing companies, virtual enterprises use local and specialization advantages of carefully selected suppliers, and themselves focus on only a few core business functions. IT management plays a key role in coordinating the activities of each player in this sophisticated weblike structure.

IT will play a key coordination role in these weblike structures

In the long run, conventional manufacturing enterprises will only be able to counter the increasing competitive pressure from "virtual" arrangements of this kind by aligning their value-added processes individually to competitive standards, and fully utilizing the benefits of internal integration. The fact that superior integrated application systems are not used sufficiently in individual core processes indicates that substantial performance reserves still remain unused.

Especially for large corporations, the concept of the virtual enterprise

COMPANY ORGANIZATION STRUCTURE

ASEA BROWN BOVERI

ABB sees itself as a network of independent local profit centers. Management is based on a sophisticated standardized management information system (ABACUS) that is used to plan and monitor the business objectives of the various divisions. Cross-country teams coordinate divisional development and manufacturing standards between the regions, and assign specific tasks to the local companies. Sophisticated IT systems are used for coordinating product development and production logistics processes internationally. The procurement activities of more than 3,500 purchasing departments are also coordinated internationally across divisions.

could be a real breakthrough in leveraging company performance. They could learn how to manage horizontally or vertically integrated business units decentrally while using the synergies of an IT-supported network without sacrificing the identity or ownership of operational functions.

IT AS A CATALYST FOR RESTRUCTURING THE INDUSTRIAL SUPPLY CHAIN

Performance can also significantly be improved across industries by restructuring the supply chain from the supplier to the end user. Our research underlined the importance of better coordination between suppliers, manufacturers, and customers via IT. This takes that idea to a new level by applying it to multistage value-added chains. The idea is not so much to accelerate data exchange or reduce communication costs with new technologies—although these objectives do determine the change process at the beginning. The primary objective is the integrated cross-industry reorganization of the communication and transaction processes.

How quickly and to what extent this concept will be accepted in an industry sector depends on a variety of factors. Innovation concepts can only be realized when the differing interests of the partners are reconciled. The speed and dimension of this development can only be predicted in specific cases. The U.S. automobile industry is playing a pioneering role, as the possibilities of ANX illustrate. Optimization of the industrial supply chain through an integrated IT infrastructure offers many benefits. The often high transaction costs between business partners can be reduced to the level of internal costs, and the service can be used whenever required. The production of goods and services is accelerated and becomes more flexible, and more creative individualized product development is possible. In addition, business relationships with international partners can be used much more effectively. Due to the plunging costs of knowledge exchange, the economics of global product development, manufacturing, and sales networks are undergoing a transformation.

By reshaping partner relations, the distribution of power in the value chain will also shift, and relationships between the business partners will be intensified. Business partners even have easier access to proprietary knowledge. The individual members will have to decide on what their contribution to the infrastructure will be, and standards and rules governing liability must be defined.

THE AUTOMOTIVE NETWORK EXCHANGE AS MODEL FOR THE FUTURE

In 1997, Chrysler, Ford, and General Motors jointly established a pilot system for an integrated private communication network that links the three automobile manufacturers to their main suppliers. Their intention is to achieve cost savings in the total supply chain of several billion dollars per year, thereby substantially reducing the manufacturing costs of all three players. The ANX network is designed to become the uniform transmission platform for all orders, including delivery data, financial data, and CAD data.

So far, first-tier suppliers are benefiting considerably from ANX. For example, Johnson Controls reduced its communication costs by 70 percent because 50 to 100 direct lines to the three automobile manufacturers were replaced by ANX.

The vision underlying ANX is even more ambitious. In the future ANX will no longer rely on paper-based processes: all the manufacturers will adjust their internal processes to ANX. This will accelerate data transmission and improve the quality of data. Today the internal transmission process of delivery data from the plant via fax until PC input via EDI takes one week. In the total supply chain, this effect is of course multiplied for each subcontractor. Johnson Controls expects considerable cost savings when subcontractors of the second and third tiers are integrated into the network.

The speed with which an industry is willing to reshape the supply chain and the extent of these adjustments primarily depend not on technological progress but on prevailing market conditions and the competitive situation in its sector. The IT-driven changes are likely to accelerate when a high innovation rate paired with short-lived product differentiation drives flexible production and shorter delivery times, or if the value chain is dominated by a few suppliers who will gain great advantage from the reorganization.

New IT technologies will drive the integrated cross-industry reorganization of communication and transaction processes

The clearest examples of IT-based reorganization of the supply chain are found, not surprisingly, in globalized industries such as the automotive and computer sectors.

MULTIMEDIA SALES AND SERVICE CENTERS AS A DIRECT LINK TO THE END USER

Why should the IT-driven optimization of value chains end at primary customers? Why should end users not be included?

Customers can already place orders on-line by phone, fax, or PC. They can get information over the Internet about the offers of alternative suppliers or share experiences in so-called communities of interest. Due to continuously declining communication costs, greater bandwidths for data transmission (e.g., ISDN), increasing private use of PCs, and safe methods of digital payment (e.g., digital cash, trust centers), the digital distribution channel will become much more important for private end users (electronic commerce). Special networks with high bandwidth for data transmission for customers from industry and trade will make it possible to establish multimedia sales and service centers in every region at reasonable costs, with specially trained staff able to provide excellent consulting services.

By combining systems for graphical product configuration and cost-based pricing with a multimedia sales interface, the digital distribution channel will become more and more attractive, offering manufacturers new opportunities for differentiation. There are many arguments in favor of making product configurators directly available to end users. Instead of selecting from product catalogues, customers can compose their preferred version of the product from different components on-line, ask for a print-out of a binding cost estimate, and place the order. The U.S. computer company Dell is the most successful Internet dealer, selling individually configurated PCs with a daily volume of more than $1 million.

A special benefit for suppliers is that customer preferences and buying behavior can be tracked and analyzed centrally. This allows on-line presentations to be adjusted to changing demand. In trade, these methods are already being used successfully. Product selection, advertising, pricing, and discounts can gradually be modified according to the needs of individual customers (the so-called segment of one).

Given appropriate network bandwidth, even elaborate multimedia product presentations and three-dimensional product simulations (virtual reality applications) can be offered via the digital distribution channel. However, multimedia applications of this kind are only likely in the medium to long term due to their high costs. Early examples illustrate this trend toward multimedia sales techniques.

Virtual reality applications make it possible to address target groups selectively

In manufacturing companies, the new information technologies can fundamentally change the interface to customers. Digital sales over the

SMART CAR: VIRTUAL PRODUCTS AT THE POINT OF SALE

DaimlerChrysler intends to break new ground with its "Smart Car" promotion. A mobile virtual reality system has been developed at a cost of around $560,000. It largely consists of a frame, steering equipment with navigation sensors, a liquid crystal screen, and a powerful graphics computer. The customer can select the color and accessories, and then the computer configures the car. The customer can experience it by means of the three-dimensional system as if the car were actually in the showroom. Plans are to install the system in car dealer showrooms from 1999.

Internet will offer even small and medium-sized manufacturing companies easier access to global markets, and it enables large corporations to bypass their traditional retailers. Multimedia, simulation software, and database marketing improve consulting and service, and make it possible to address target groups selectively.

IT stars will systematically use the potential of these new techniques to strengthen their market position and increase their lead.

NEW CHALLENGES FOR IT MANAGEMENT

We have already examined the many factors driving the complexity of IT management: the productivity gains of new software technologies, rapidly expanding areas of application, and the pressure to adapt/migrate existing systems. In many companies IT management is already overwhelmed by the fundamental issues that need resolving. Coping with this increase in complexity presents a genuine challenge.

To stand this test and avoid huge performance gaps compared to the best in their industry sector, companies with average or poor IT management performance to date must reshape their approach. Three main challenges are emerging:

INCREASING DEMANDS ON IT PLANNING

Only best-practice companies have so far managed to install farsighted, business-oriented planning of their IT budgets. Planning is reactive rather than proactive across all industry sectors: insufficient use is made of reliable planning tools, and cost data are often intransparent. The trends already described will increase the demands of IT planning and greatly expand its spectrum of application.

At the same time, the complexity and variety of the systems architectures to be planned are growing. In addition to the conventional systems landscape, new technology platforms are emerging that need to be integrated. Combining conventional integrated application systems with intranet applications will require especially careful planning to ensure optimal configuration. In the medium term, companies must be prepared for substantial migration and modification costs due to these factors.

Usually investments in applications and base systems cannot be planned independently from one another. For example, the benefits of a new database system as a systems platform are dependent on the IT applications for which it is designed. The economic evaluation of IT investments, a difficult enough task in the past, will become even more complex as the networking of systems and applications grows. Against this background, the key task will be to build a modular IT organization with a strong, independent, top management–oriented IT planning and consulting group as discussed in Rule 6.

DEVELOPMENT OF NEW SOLUTIONS FOR SYSTEMS OPERATION AND SERVICE

The IT operations of many companies are already functioning inefficiently when you consider the insufficient user involvement and lack of data on real cost-benefit ratios. Systems architecture will become even more complex and trouble pronc as application systems are distributed flexibly throughout the entire network. Network management and systems operations will face a severe challenge to ensure stable network operations companywide across all locations.

A solution could be outsourcing to high-performance, specialized providers—from mainframe activities and network services through to complete configuration and service of PC systems. The competitiveness of internal IT services will therefore have to be compared carefully to external service providers. If the company decides to outsource specific activities, partner management will be very important. Without installing professional partner management, companies will be unable to capture the service and productivity benefits they expect on a sustainable basis.

SHIFTING ROLES IN IMPLEMENTATION MANAGEMENT

Besides the replacement of outdated systems, the key objective of new systems implementation should be to integrate and supplement existing operational systems, both within the company and in association with other stakeholders, from suppliers to customers. For new areas of application, companies will, as an add-on to their PPS system, increasingly use dedicated component software with a high degree of integration (e.g., for supply chain management). As described in Rule 7, the development of proprietary systems for these purposes needs to be restricted to areas with a clear strategic advantage. New configuration tools will greatly assist the parameterization of existing software and reduce the costs of implementation support.

However, the demands on implementation management will continue to be high. The migration of existing application systems and data will involve considerable effort and expense, frequently caused by insufficient documentation, but also by the difficulty of integrating proprietary systems. The complexity of interface and integration management will rise as the variety of applications and degree of systems integration required grow. Successful implementation of new user systems will scarcely be possible without first conducting a professional business process analysis. The tasks of implementation management will thus continue to shift from the adaptation of software to the redesign of processes.

SUMMARY

As ever, the future holds both opportunity and threat. On the one hand, new technology will continue to enable IT stars to make quantum leaps in effectiveness. On the other, poor management of IT can result in a cost explosion. Each of the four IT cultures can benefit from internalizing the seven rules for managing superior information technology performance that we have identified, but the road to improvement will take a different course for each of the groups (see Exhibit IV–4).

We believe that a laggard can become a star in two to three years, but only if the company rethinks its notion of IT. Instead of being regarded as a limited specialist task, IT must become a key concern of top management. No longer simply a means of reducing manufacturing costs through automation, it must be seen as a tool for optimizing almost any business process. While no manufacturing operation can guarantee to transform itself simply by following a set of guidelines, there is ample evidence to suggest that the journey from IT laggard to IT star will make a striking difference to a company's bottom line.

Exhibit IV–4
Routes to Improvement of IT Performance, by Type of IT Culture

Effectiveness of IT ↑		
	Big IT spenders • Standardization of IT • IT cost control • Intelligent IT outsourcing	**IT stars** • Review IT support for business processes
	IT laggards • Top management involvement • Standardization of IT • Planning/control of IT projects	**Cautious IT spenders** • Top management involvement • IT as internal service provider • IT support for business processes
	Efficiency of IT →	

Source: McKinsey & Company, Inc.

APPENDIX A

ABOUT THE EMPIRICAL RESEARCH

This appendix explains the approach and scope of the McKinsey research on IT performance in the manufacturing sector.

APPROACH

From July 1996 to September 1997, in cooperation with the Institute of Production Engineering and Machine Tools at Darmstadt University of Technology, Germany, we carried out an international survey in the manufacturing sector. The survey included an optimal mix of research methods:

- We analyzed 102 leading companies from many relevant manufacturing industries. Of these, 72 took part in the full survey (see Exhibit A–1). In-depth questionnaire-based interviews were conducted with nine top managers at each participating company. Around 3,500 items of data were gathered on each organization.
- Thirty companies were interviewed on specific subjects (e.g., outsourcing, rapid prototyping, product configuration, and integrated standard software). Discussions were also held with selected suppliers of these IT services and products.

SCOPE OF THE SURVEY

The sample offered a well-balanced structure with regard to breakdown by geographic region, type of industry, and size of company (see Exhibit A–1).

Exhibit A–1
Distribution of Companies Surveyed, by Region, Industry, and Size

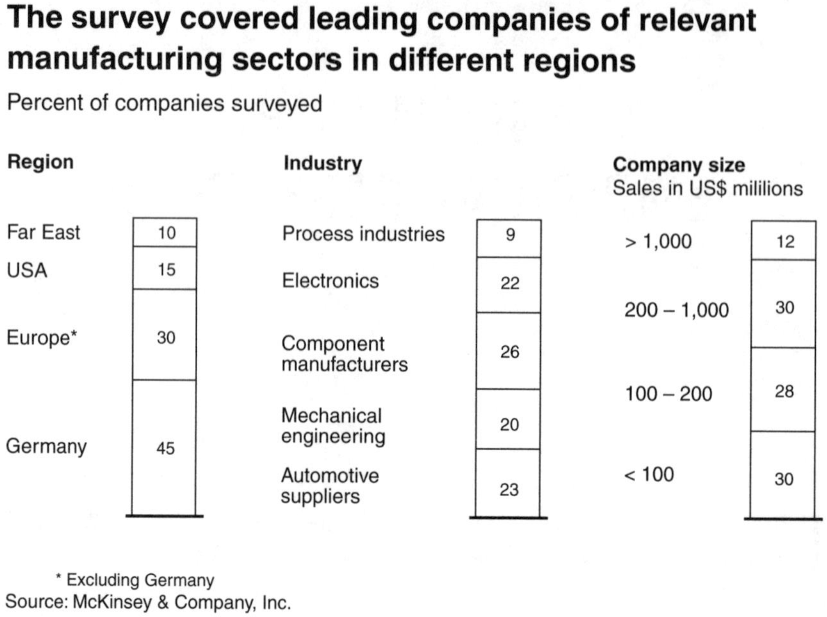

The survey covered leading companies of relevant manufacturing sectors in different regions

Percent of companies surveyed

Region	
Far East	10
USA	15
Europe*	30
Germany	45

Industry	
Process industries	9
Electronics	22
Component manufacturers	26
Mechanical engineering	20
Automotive suppliers	23

Company size Sales in US$ mililions	
> 1,000	12
200 – 1,000	30
100 – 200	28
< 100	30

* Excluding Germany
Source: McKinsey & Company, Inc.

The companies were located in Europe, the United States, and Asia. The manufacturing industries ranged from automotive suppliers, mechanical engineering companies, component manufacturers, and electronics firms to companies in the process industry. Operating units with revenues from $30 million to $9 billion were analyzed. Most of the smaller units are affiliates of major corporations with decentralized IT organization structures. Seventy-two percent of the respondents belong to industrial groups with revenues of over $1 billion.

Appendix B explains the methodology of the newly developed McKinsey system for measuring IT performance.

APPENDIX B

MEASURING THE PERFORMANCE OF IT

To measure the performance of IT and its impact, we developed a system that uses indicators of corporate success, business process performance, and the performance of IT along the dimensions of efficiency and effectiveness.

The impact of IT management on corporate success was measured directly. The indicators of business process performance assess IT performance in specific core processes such as product development, marketing/sales, order processing, and service. Here the contribution of IT to corporate success is indirect, but the process performance itself has a direct impact on corporate success (for further details on this, see Part I). Care was taken to define independent indicators for corporate success, business process performance, and IT performance.

INDICATOR OF CORPORATE SUCCESS

The indicator of corporate success (see Exhibit B–1) was calculated on the basis of four criteria: return on sales, change in return on sales, change in sales, and change in market share. The importance of these criteria was rated equally. The analysis covered the period from 1993 to 1996.

Exhibit B–1
Indicator of Corporate Success

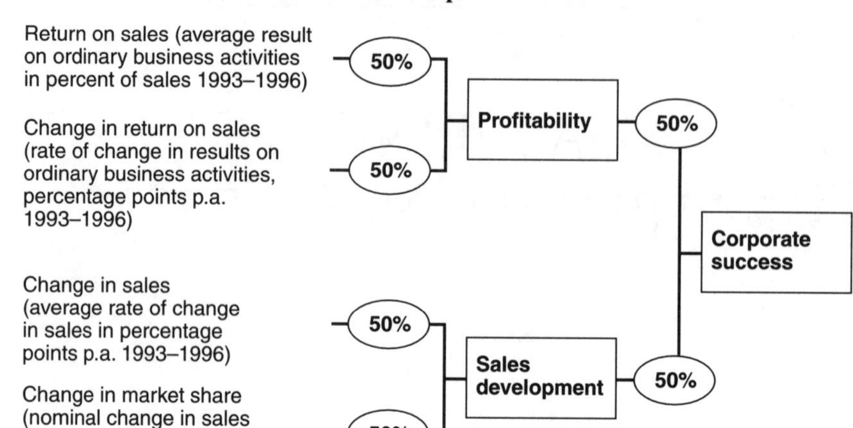

Source: McKinsey & Company, Inc.

INDICATORS OF BUSINESS PROCESS PERFORMANCE

INDICATOR OF PRODUCT DEVELOPMENT PERFORMANCE

The method used to calculate the indicator of product development is described in Exhibit B–2.

We developed an indicator that considers three variables—time, quality, and costs of product development—with the importance of each variable rated equally. The quality of product development was evaluated on the basis of two variables of equal weight: rate of product modifications and percentage of sales of superior products.

INDICATORS OF OPERATIONAL CORE PROCESS PERFORMANCE

Indicators were calculated for each of the following operational core processes: marketing/sales, order processing, and after-sales service (see Exhibit B–3). Each of these indicators considers two to three variables that are equally weighted.

Exhibit B–2
Indicator of Product Development Performance

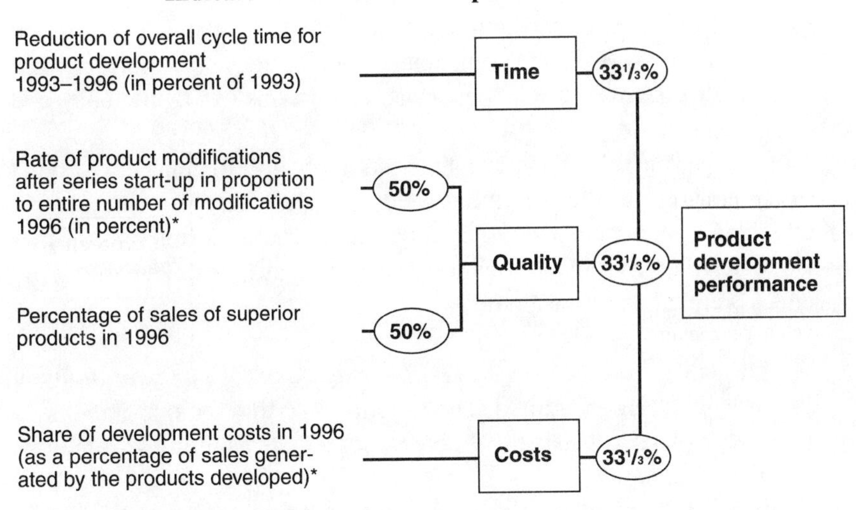

* The objective should be to minimize this percentage.
Source: McKinsey & Company, Inc.

Exhibit B–3
Indicators of Operational Core Process Performance

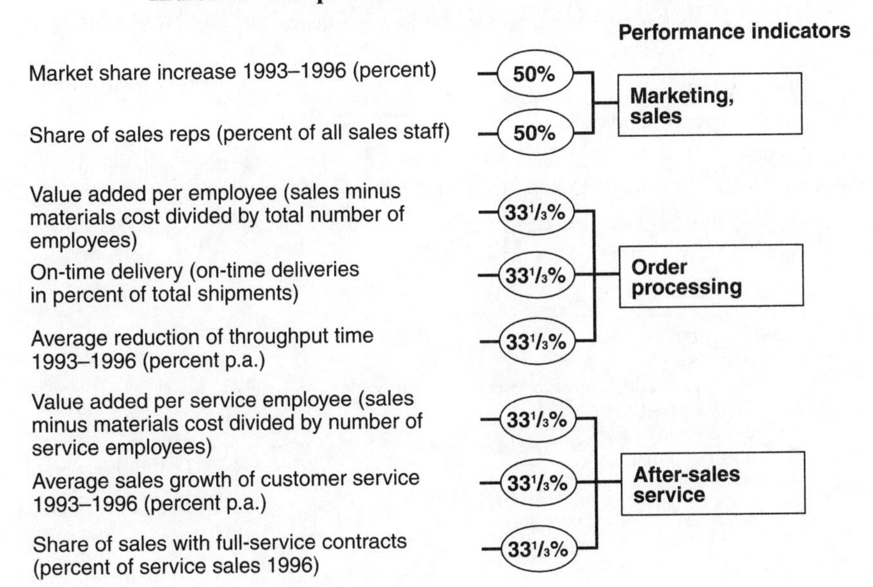

Source: McKinsey & Company, Inc.

INDICATORS OF IT PERFORMANCE

The method for calculating the indicators of IT performance is described in Exhibit B–4. This is a sophisticated two-dimensional performance measurement system rating IT performance in terms of IT efficiency and effectiveness.

For measuring IT efficiency, IT costs and the efficiency of project management were evaluated, rating each variable with equal weight. The IT effectiveness in each core process was measured in terms of the functionality, availability, and utilization rate of IT systems. The following variables were taken into account:

- Functionality—the *number of IT-supported functions or operations* in the process being examined. The functionality that the respective systems offer plays a decisive role here rather than whether these functions are actually used.
- Availability—the availability of information for users. This is dependent on the *availability of IT systems* (i.e., low failure rate, quick response time, number of terminals/PCs per user), the *quality of data* provided (i.e., consistent, complete, error free), and the *degree of integration.* As far as the degree of integration is concerned, *internal* inte-

Exhibit B–4
Indicators of IT Performance

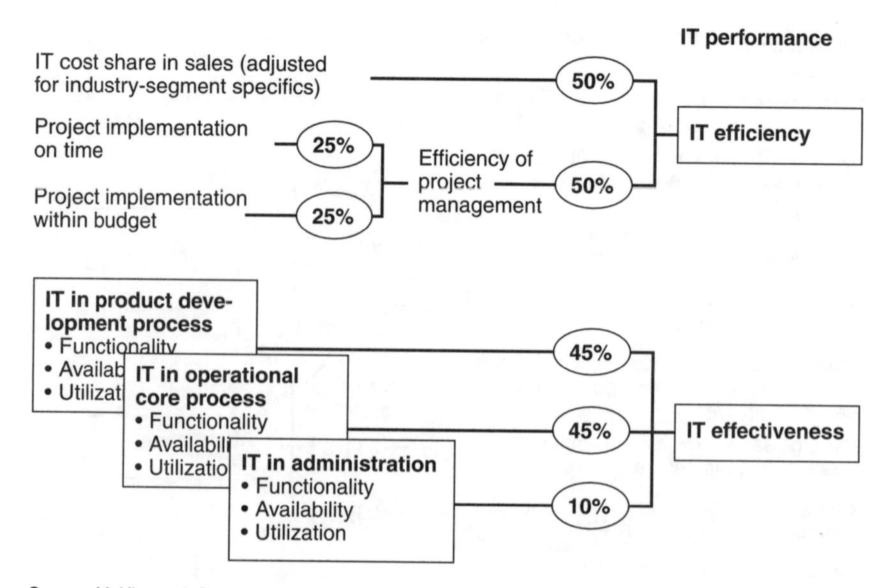

Source: McKinsey & Company, Inc.

gration of IT systems (i.e., availability across business functions) was distinguished from *external* availability of information (i.e., for development partners, suppliers, and customers).

- Utilization rate—an assessment of the extent to which the functions provided are used, whether the utilization aims at improving total process performance, as well as the acceptance, satisfaction, and training level of the IT users.

Exhibit B–5 illustrates an example of the indicators used to evaluate IT effectiveness.

Exhibit B–5
IT Effectiveness Indicator: Examples

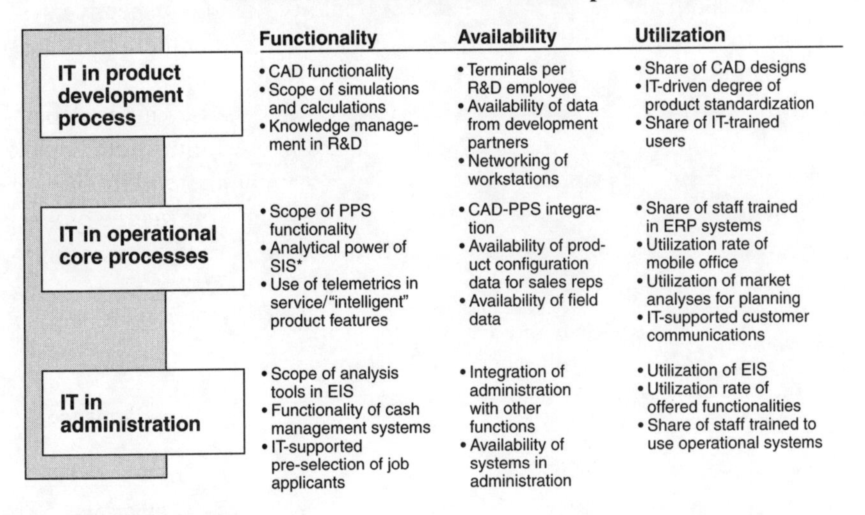

	Functionality	Availability	Utilization
IT in product development process	• CAD functionality • Scope of simulations and calculations • Knowledge management in R&D	• Terminals per R&D employee • Availability of data from development partners • Networking of workstations	• Share of CAD designs • IT-driven degree of product standardization • Share of IT-trained users
IT in operational core processes	• Scope of PPS functionality • Analytical power of SIS* • Use of telemetrics in service/"intelligent" product features	• CAD-PPS integration • Availability of product configuration data for sales reps • Availability of field data	• Share of staff trained in ERP systems • Utilization rate of mobile office • Utilization of market analyses for planning • IT-supported customer communications
IT in administration	• Scope of analysis tools in EIS • Functionality of cash management systems • IT-supported pre-selection of job applicants	• Integration of administration with other functions • Availability of systems in administration	• Utilization of EIS • Utilization rate of offered functionalities • Share of staff trained to use operational systems

* Sales information system.
Source: McKinsey & Company, Inc.

GLOSSARY

Branching and bounding approach

This is a term from operations research for discrete problems (e.g., allocation problems) that cannot be solved using effective analytical processes or an enumeration process (due to the excessive calculations that would be needed). If the problem can be formulated as n discrete variables, each of which has k possible values, it can be solved using branching and bounding. This means dividing up and restricting the solution space to avoid complete enumeration (dividing up = branching; restricting or containing = bounding).

Business engineering workbench (BEW), also called dynamic enterprise modeling (DEM)

These are tools for simple and flexible modeling of business processes on computer. These orgware tools provide access to best-practice reference processes and functional models for business process optimization, and greatly accelerate implementation of the integrated standard software SAP R/3 (BEW) and Baan IV (DEM) via a so-called mapping function. The mapping allows direct configuration of the work flows defined in BEW/DEM in the R/3 and Baan IV base software. Experts estimate that tools like BEW/DEM reduce the implementation time for integrated standard software by 40 to 60 percent. Mapping also allows a simpler adjustment of business processes during the utilization time of the base software, supporting a continuous improvement process. The comprehensive definition of all work steps and constant checking of results ensure high quality.

Componentware, component software

Today one of the major challenges to software development is to develop new application systems faster, at reduced costs, and of better quality. The basic principle of component software for software development is to include components that already exist instead of having to program application systems from scratch. Such components are software modules that can be reused and used in different environments (i.e., hardware configurations, operating systems, and networks). Component software can be generated conveniently by means of object-oriented programming languages.

Computer-aided design (CAD)

With CAD systems, objects or products can be designed on the screen of a workstation or high-performance PC.

Computer-aided software engineering (CASE)

CASE systems offer numerous IT tools and procedures designed to improve the productivity of systems analysts and programmers substantially. Running on workstations or high-performance PCs, these are generators of application systems that allow graphic-oriented front-end automation of the software development process.

Computer-integrated manufacturing (CIM)

CIM is the generic concept for very heterogeneous IT applications to integrate all production work flows: groupware systems, information systems, factory automation, and others. All these applications are designed to connect different systems and functions to enable effective production.

Concurrent engineering (CE)

This means performing design and product development activities simultaneously rather than sequentially. This requires professional IT support and cross-functional cooperation in project teams. All relevant departments and business functions should be represented on the project team.

Data warehousing (DW)

This is a generic term for various methods of storing, retrieving, and managing large data volumes. A DW system consists of data, a meta database, and a data manager. A data warehouse is an organized collection of cross-functional data, usually broken down by time, divisions, regions, customers, and so forth. The meta database manages all specifications concerning semantics, time horizon, quality, source, and retrieval method, as well as details on possible modifications of the data model. The data manager controls the data import from application systems such as the production planning system (PPS/ERP), and also from external databases (e.g., on-line databases or the World Wide Web). A relational database system is usually the basis of a DW. The DW provides data for user software such as decision support systems and executive information systems.

Decision support system (DSS)

"Intelligent" DSS of the latest generation are interactive software tools that run on workstations or high-performance PCs. They are designed to support managers in making decisions in well-structured planning or decision environments. The concept is based on decision and game theory. DSS offer the following features as parameters for decision making: alternative scenarios, status descriptions, definition of interactions and preferences, and impact diagrams.

Dynamic enterprise modeling (DEM)

See Business engineering workbench.

Electronic data interchange (EDI)

This is paperless data transfer—in batch processing—via network or data carrier between IT systems of different companies. All types of formatted documents can be transmitted—for example, order notes, acknowledgments of order, receipts, design prototypes, and engineering and product data when working with product development partners. Prerequisites are compatibility of hardware and software configurations with standardized networks and effective security controls. Increasingly, external EDI clearing centers are being established that function as neutral intermediaries between transmitter and receiver. Above all, they perform the transition and standardization of data required for global data exchange. They also guarantee data protection and document authenticity.

Enterprise resource planning (ERP)

Integrated systems of the latest generation for production planning and scheduling, manufacturing resource planning, and other functional areas like accounting and sales management. These are "open" systems with client/server architecture. They offer graphic user interfaces (shells). Apart from standard functionalities, they also provide modules for process management, quality assurance, and regular reporting. Customized user applications are possible.

Executive information system (EIS)

These systems support top management in planning and controlling important performance indicators effectively. Unlike DSS, they are specifically tailored to the needs of top management decision makers; they are less detailed and mostly focus on financial data.

Extranet

This is an extension of an intranet that offers clearly defined and authorized user groups (e.g., suppliers, customers/dealers, distributors) limited on-line access to data in the intranet that are relevant to them.

Forecasting systems

Today expert systems (artificial intelligence) are used for complex forecasting problems (e.g., long-term development of demand or competition in mature industries). Typically, time-series forecasts are based on Markov models and sometimes neuronal nets.

Global Standard of Mobile Communication (GSM)

European standard for mobile telephones.

Internet

The largest computer network in the world, linking international and national networks such as MILnet, NSFnet, and CERN, as well as countless regional and local networks. The Internet combines networks of variable size that are interconnected by routers to a single virtual network. The basic characteristic of the Internet is the open standardized network architecture allowing any PC user to become connected to any other via TCP (file transfer protocol)/IP (Internet protocol) transmission protocols.

Intranets

In-house computer networks based on Internet technology but operated independently from it. Frequently, intranets have standard interfaces to the Internet so that users can plug into it and access Internet programs.

Java

A programming language developed for use with the Internet by Sun Microsystems. Programs written in Java can be downloaded from the Internet onto any PC, and can be used regardless of the systems platform (i.e., type of CPU or operating system of the PC).

MRP I

See Production planning and scheduling systems (PPS).

MRP II

See Production planning and scheduling systems (PPS).

Product data management (PDM)

Product data management systems run on mainframes or workstations and communicate with various software applications; they can also be linked to document management and retrieval systems. These systems work on an integrated database concept that makes all *product-related data* accessible and can also organize and save the data. A PDM tool can be used to systematically track the entire life cycle of a product. These tools provide entry restrictions and controls, store existing data links, and support control systems for definition of data flows and processes. Additionally, they provide news and memo functions.

Production planning and scheduling systems (PPS)

PPS is the generic term for older *material requirements planning* (MRP I) systems and recent *manufacturing resource planning* (MRP II) systems. PPS systems support management in effectively planning resource utilization throughout the production process. They also provide information and make scheduling proposals. Modern MRP II systems integrate business planning, production planning, work planning, capacity planning, materials purchase planning, and the corresponding operations planning. Ideally, PPS systems deliver target output figures for the operational units as well as target returns for financial planning. They can also answer simple "what-if" questions via a simulation function. The latest MRP II systems are open systems, oriented to the ERP standard.

Rapid prototyping

Prototypes of parts or products can be manufactured automatically on the basis of a three-dimensional computer-assisted design model. A well-known method is stereo lithography.

Real-time System Version 3 (R/3)

This is the current version of the integrated standard software of SAP. It is based on client/server architecture. R/3 can be used on PCs with a graphic user interface, usually on Windows or Windows NT. Transactions are controlled on UNIX servers in a local-area network environment.

Simulation

Products or parts can be calculated, tested, and optimized under different conditions in the product development process using simulation models. Search algorithms or stochastic methods are used to find an optimal constellation of parameters. As search algorithms, methods analogous to nature are used such as evolutionary algorithms or simulated annealing and, recently, neuronal nets as well as taboo search.

Virtual prototyping (VP)

VP enables creation of a digital prototype of a product or part that is to be developed. First, functional tests can be conducted on the digital prototype under conditions of virtual reality. In addition to the cost savings achieved by using physical prototypes less, VP greatly reduces product development cycle times.

Virtual reality (VR)

This means an audiovisual reality created by computer simulation. CAD applications are best known in technical design and architecture. A headset—an integral helmet that is connected to a computer—allows three-dimensional effects created through two independent interior displays, each of which provides pictures for one eye. A so-called data glove substitutes for the keyboard.

Work flow management systems (WMS)

These are software applications that run on mainframes or client/server architectures. They control and support the different steps of a work flow. Unlike object-centered application systems, WMS exactly orient themselves to the underlying business process. The computer representation is based on a specific work flow logic. First, the individual work steps of a business process—as far as they can be technically defined in IT terms—are modeled graphically. The end product is an exactly defined, complete and consistent work flow that is programmable. In a second step, the work flow is executed—interpreted with regard to the organization structure and structure of operations involved, digitally configured, and provided to authorized employees in the form of operating lists. Groupware systems such as Lotus Notes are among the best known WMS.

World Wide Web (WWW)

This is a worldwide public multimedia information network within the Internet. The graphic user interface is generated using the standardized page description language HTML (Hypertext Markup Language). On these interfaces, data are organized in knots or buttons that can be connected through pointers displayed as pictograms, for example, or selection menus. HTML allows navigation through web sites via hyperlinks.

INDEX

ABOUT THE AUTHORS

The McKinsey team of authors consists of

Ralf Augustin, Darmstadt
Engineer, group head of Corporate Strategy, Institute of Production Engineering and Machine Tools at Darmstadt University of Technology; consultant at McKinsey & Company, Inc. (since June 1998)

Dr. Günter Bulk, Düsseldorf
Quality leader of GE Capital Information Technology Solutions (Europe); former team leader, member of the Information Technology/Systems Practice and Assembly Industry Sector

Christopher Höfener, Düsseldorf
Physicist, consultant, project manager of the "Do IT Smart" research project, member of the Assembly Industry Sector

Dr. Rolf-Dieter Kempis, Düsseldorf
Engineer, director, member of the worldwide Leadership Groups for the Automotive and Assembly Sector, as well as Steel Industry

Dr. Jürgen Ringbeck, Düsseldorf
Mathematician and business economist, principal, head of the Transportation Sector in Germany, member of the worldwide leadership groups for Information Technology and Growth Strategy

Berthold Trenkel-Bögle, Seoul/Tokyo
Engineer, M.B.A., team leader, member of the Transportation, Telecommunications, IT, and Media Sector